D0897648

Innovative Control Charting

Also available from ASQ Quality Press

Glossary and Tables for Statistical Quality Control, Third Edition
ASQC Statistics Division

Statistical Process Control Methods for Long and Short Runs, Second Edition
Gary K. Griffith

SPC Essentials and Productivity Improvement: A Manufacturing Approach
William A. Levinson and Frank Tumbelty

Managing the Metrology System, Second Edition
C. Robert Pennella

Practical Product Assurance Management
John Bieda

To request a complimentary catalog of publications, call 800-248-1946.

Innovative Control Charting

Practical SPC Solutions for Today's Manufacturing Environment

Stephen A. Wise
Douglas C. Fair

ASQ Quality Press
Milwaukee, Wisconsin

Innovative Control Charting
Practical SPC Solutions for Today's Manufacturing Environment
Stephen A. Wise and Douglas C. Fair

Library of Congress Cataloging-in-Publication Data
Wise, Stephen A., 1960–
Innovative control charting: practical SPC solutions for today's manufacturing environment / Stephen A. Wise, Douglas C. Fair.
p. cm.
Includes bibliographical references and index.
ISBN 0-87389-385-9 (alk. paper)
1. Process control—Statistical methods. 2. Production management. 3. Charts, diagrams, etc. I. Fair, Douglas C., 1964– . II. Title.
TS156.8.W58 1997
658.5'62'015195—dc21 97-21281
CIP

©1998 by ASQ
All rights reserved. No part of this book may be reproduced in any form or by any means, electronic, mechanical, photocopying, recording, or otherwise, without the prior written permission of the publisher.

Trademark Acknowledgment
Trademarks owned by International Quality Institute, Inc. Used with permission.

10 9 8 7 6 5 4 3

ISBN 0-87389-385-9

Acquisitions Editor: Roger Holloway
Project Editor: Jeanne W. Bohn

ASQ Mission: To facilitate continuous improvement and increase customer satisfaction by identifying, communicating, and promoting the use of quality principles, concepts, and technologies; and thereby be recognized throughout the world as the leading authority on, and champion for, quality.

Attention: Schools and Corporations
ASQ Quality Press books, videotapes, audiotapes, and software are available at quantity discounts with bulk purchases for business, educational, or instructional use. For information, please contact ASQ Quality Press at 800-248-1946, or write to ASQ Quality Press, P.O. Box 3005, Milwaukee, WI 53201-3005.

For a free copy of the ASQ Quality Press Publications Catalog, including ASQ membership information, call 800-248-1946.

Printed in the United States of America

 Printed on acid-free paper

American Society for Quality
ASQ®

Quality Press
611 East Wisconsin Avenue
Milwaukee, Wisconsin 53202
Call toll free 800-248-1946
www.asq.org
http://qualitypress.asq.org
http://standardsgroup.asq.org
http://e-standards.asq.org
E-mail: authors@asq.org

To my darling wife, Joanna, and
action-packed daughter, Aubrey Sue.

Steve Wise

To the memory of my very special mother and friend,
Shirley Chism Fair.

Doug Fair

Contents

Preface

"It's a great idea in theory, but my process is different. SPC just won't work in my situation." Again and again, this was the response from Boeing Company engineers, managers, and machine operators. As degreed statisticians, it was frustrating for us to learn that, in many cases, they were right. Fresh out of college and idealistic about applying statistical process control (SPC), we visited shop floors and found, to our dismay, that the Boeing manufacturing operations were the equivalent of an enormous, complex job shop.

Huge numbers of different parts, configurations, and jobs seemed to come across each manufacturing area. Repeat jobs seemed to never come back to the same machine. Machining centers or processes that continuously manufactured a single product or part number were rare indeed. In other words, it was difficult to find an area where traditional control charts could be used to benefit those who would be using them. Briefly, we thought that maybe the Boeing employees were right. Maybe their situations were too different. In the beginning, it seemed so.

But as we searched for alternative SPC methods, we uncovered many different techniques that could allow us to apply control charts in Boeing's job shop atmosphere. However, our approach to applying SPC would have to be creative and flexible. Above all else, the control charts would have to be useful to the manufacturing managers, engineers, operators, statisticians, and quality assurance professionals who would be using SPC on a daily basis.

In keeping with our early work at Boeing, this book has been specifically written for the SPC users just identified. Our primary audience is those people who have the daunting task of understanding and implementing SPC in today's complex manufacturing situations.

The purpose of this book is to provide solutions for using control charts in almost any manufacturing process. This book is intended for people who face the challenges of small production runs, multiple characteristics, limited data collection opportunities, sophisticated customers, and demanding manufacturing situations.

Even though this was our primary focus, *Innovative Control Charting* has also been written for those who need to apply SPC in traditional, long production run environments as well. Also, those people who have had no exposure to SPC should find this to be a useful text. For the uninitiated, we have concentrated on making the wording clear, concise, and easily understandable. So you need not have a degree in statistics to understand the information contained within.

For those unfamiliar with SPC, the first few chapters deal with the essentials. Chapters 2 through 10 cover the basics of SPC, control chart theories, process capability studies, sampling and data collection strategies, and the three traditional control charts. These chapters form the foundation upon which the rest of the book has been written. Chapters 11 through 30 concentrate on charts that describe creative ways of dealing with the limitations posed by traditional control charts.

In addition, this book has been written with manufacturing in mind. Specifically, it deals only with applications and control charts for variables (measurements) data. In turn, there will be no discussion of service industries or attributes data applications.

Finally, *Innovative Control Charting* has been written with practicality in mind. The text has been minimized, and there appear only short descriptions and examples of traditional and nontraditional control charts. Our intent was to provide a handbook of control charting tools that people could use to quickly identify and understand the chart needed to solve their SPC challenge. Given this intent, the reader will find the text de-emphasizes theory, probability, distributions, and formula derivations. A large number of excellent texts already exist that include this information and, therefore, will not be covered in this book. Instead, we wanted to make this the kind of book that would get dog-eared and dirty from an overabundance of use by the quality professionals whose task it is to successfully implement SPC.

Acknowledgments

We would like to acknowledge and thank the many people who have contributed both directly and indirectly to the writing and completion of this book. First we would like to thank Joanna and Jody (our spouses) and Aubrey Sue; Douglas, Jr.; and Connor (our children) who have been so extraordinarily patient and supportive of us during the exciting, yet time-consuming process of writing this book. A special thank you to Jody, for giving me the time needed to write, for making the coffee, for taking care of the kids, and for providing to me her incredibly supportive and positive attitude over the approximately four and a half years it has required to bring this work to completion.

We would also like to thank Jerry Sprogue and Harvey Hailer of the Boeing Company for their leadership and encouragement and for giving us a chance. Jerry and Harvey gave us the original opportunity to work together, research, and prove many of the techniques contained in this book. Many thanks go to Harvey for his idea to write a book in the first place. We would also like to thank the Boeing Company and its many special employees who pushed us to come up with innovative statistical process control (SPC) applications for their diverse manufacturing challenges. Additionally, we would like to thank the University of Tennessee in Knoxville and all of the faculty members in the Department of Statistics for providing us an excellent learning environment and a superior education.

A special thank you goes to Davis Bothe and Chad Cullen of the International Quality Institute. Davis originally introduced us to the wonderful world of creative process control statistics. Davis' pioneering ideas have removed the excuses. We would also like to thank Dr. H. Alan Lasater for his thorough examination of our book and his many suggestions on how to improve it.

Another thank you is appropriate for Michael Lyle and Chris Kearsley of Lyle-Kearsley Systems. They have devoted their talents to creating the very best SPC software on the market, *InfinitySPC*. Most control charts and histograms found in this text were created using *Infinity*.

Last, we would like to thank all the people who stopped us and said, "What we would *really* like to be able to do with SPC is. . . ." Those people will never know the thrill they have given us over the years.

Stephen A. Wise
Douglas C. Fair

Chapter 1

Introduction

While in the throes of a deep recession in the early 1980s, the Big Three automakers struggled to increase profitability and market share. How was it that the Japanese auto manufacturers had made such significant strides in productivity and quality? Were their cars really that good? If their quality was better, how were such high levels achieved? Speculation, at first, centered on the Japanese work ethic and attitude as well as a much publicized use of consensus decision making and worker involvement.

However, what fascinated many business leaders in the West was the Japanese companies' use of statistical process control (SPC). It was claimed that SPC helped to reduce scrap and rework and simultaneously improve reliability and quality for manufactured products. The word spread quickly and the use of statistical tools soon became the rage in companies throughout the western world.

More than a decade has passed since SPC was thus blessed as a major contributor to continuous improvement and total quality management (TQM) philosophies. Since then, many other tools, techniques, and philosophies have been introduced to slash manufacturing costs and boost productivity. Among them are organizational development and team involvement techniques, quality function deployment (QFD), theory of constraints, kaizen, reengineering, lean manufacturing, and just-in-time (JIT) inventory systems, to name a few.

Many of these techniques proved to be so overwhelmingly successful that, in the late 1980s and early 1990s, SPC lost some of its luster. Though still used by many companies in their search for improved efficiencies and operating costs, SPC fell out of grace as the darling of the quality improvement movement. This was especially evident when companies tried to use both SPC and JIT in their operations. Because of the interest in decreasing lot sizes and short production runs, JIT philosophies seemed to conflict with SPC.

Today, the focus on SPC has diminished even more because customers have become more demanding. Not only do they want more features and options in their products, but customers also demand more product customization. More options and customizations have translated into an ever-growing list of part numbers, manufacturing complexity, and a geometric progression of critical dimensions and characteristics requiring control. In short, the advent of small lot sizes, increased part complexity, and customization has made it exceedingly difficult to effectively use traditional SPC.

The result of these industry developments has highlighted three primary limitations faced by those using traditional SPC techniques.

1. Fifteen to 25 data points are needed before calculating statistical control limits.
2. Only one characteristic can be tracked per control chart.
3. Even if part characteristics are similar, other differences such as material type or specification limits necessitate separate control charts.

Limitation 1: 15 to 25 Data Points Are Needed Before Calculating Statistical Control Limits

Traditional SPC techniques were developed to be applied to long, rarely changing production runs. To develop a control chart that accurately reflects process performance, one traditionally needs between 15 and 25 data points. Henry Ford was once alleged to have said that his customers could buy any color Ford vehicle they wished, as long as it was black. Black quickly became the color of choice. If one wanted to develop a control chart to track the thickness of black paint, no problem. No problem because 15 or 25 cars might be painted in the blink of an eye. It would have been easy (if the technology were available to obtain paint thickness data) to develop a control chart because the opportunities for data collection would have been plentiful.

However, no auto manufacturer in today's marketplace would dare offer a vehicle with only one color choice. In fact, the amount of customization and options made available by auto manufacturers is staggering. Short production runs and small batch sizes are what's required to cater to customers' needs. Short production runs result in numerous process changes, which, in turn, translate into smaller amounts of data for a given production setup.

In this environment of enormous part numbers, gathering 15 to 25 data points from ever-decreasing lot sizes almost becomes an impossibility, as does the development of a traditional control chart. After all, if one is going to make only a handful of parts, how can control chart development be justified? And even if it could be justified, how would it be done? These are indeed difficult questions to answer if one is armed only with traditional SPC techniques.

Limitation 2: Only One Characteristic Can Be Tracked per Control Chart

This is really not a limitation if there are just a handful of important characteristics to control. Once again, though, customer demands have translated into tremendous numbers of important characteristics to track. We have seen a situation in which engineers had identified 72 characteristics to be controlled with control charts. Each characteristic had an $\overline{X}$ and range (R) chart associated with it. This meant there were 144 different graphs to maintain.

The enormous quantity of data necessitated a separate meeting room in which to hang all the control charts. The result was data overload. It was extremely difficult to derive any meaningful information out of all those graphs. The amazing thing was that those 72 different characteristics, 144 graphs, and dedicated display room were for a *single* part. The data represented only one of hundreds of items manufactured at the facility.

Sometimes even the same characteristic on the same part might require separate charts. For example, take a twin spindle mill that is facing a stainless steel surface to a single specified thickness. If the left side of the part is milled by spindle A and the right side is milled by spindle B, one would need two separate control charts. The logic is simple: try to identify assignable causes that might be present at one spindle but not at the other. Spindle A may have new inserts whereas spindle B's might be worn. Or, the coolant directed at the workpiece under spindle A might have greater pressure than spindle B. The challenge is to determine what effect the different tools had on the thickness of the stainless steel part.

In effect, the operator would want to ensure that the thickness is uniform across the piece. Therefore, a control chart would be developed to evaluate the consistency of the left side of the milled surface (milled by spindle A) and another chart would be established for the right side (milled by spindle B).

Multiple characteristics result in multiple charts, which translates into more work for the operator and less attention to his or her manufacturing duties. The point is that traditional SPC techniques are not very efficient when one needs to evaluate multiple characteristics.

Limitation 3: Even If Part Characteristics Are Similar, Other Differences Such as Material Type or Specification Limits Necessitate Separate Control Charts

Take, for example, the paint discussion earlier. Say a manufacturer is interested in developing a control chart for paint thickness of two types of paint: flat white and metallic green. Just take paint thickness measurements regardless of paint type and put them on one chart, right? After all, paint is paint, right? Well, not exactly. Each type of paint may have a different target specification for thickness. One may be required to be thicker or thinner than the other. Because of this, one must set up two different control charts because of the difference in target paint thicknesses. One chart would be needed to track the thickness of the metallic green and another for the flat white. Nine different paint colors would necessitate nine different control charts.

Even if the parts or processes are similar, they are still different and logically might behave in dissimilar fashions. Washers of differing thickness, bored holes of different diameters, and hydraulic pumps with different performance requirements all produce a realistic need for separate control charts. This limitation is similar to limitation 2 because the result is more control charts, more complexity, and more hassle for people using SPC.

The three limitations described here have bedeviled SPC implementers for years. For the reasons outlined, many people concluded that it was a waste of time to make control charts. Nothing could be further from the truth. The limitations here are not the result of the failure of traditional SPC techniques—but rather the need for a different, nontraditional set of charting methods.

Traditional techniques are still appropriate and effective when applied to long production run processes. However, using the traditional $\overline{X}$ and *range chart* for a lot size of eight pieces with five different characteristics is a failure in the use of the chart. The $\overline{X}$ and *range chart* is still a good tool, it is just being applied in the wrong situation. This example might be likened to using a hammer to drive dry wall screws. You can do it, it's just that it is difficult and the result is very messy. It doesn't mean that the hammer is a bad tool. It does, however, highlight the fact that another tool might be more effective and efficient in accomplishing the task.

The same is true with control charts. One only needs the right control chart tool to deal with the complex manufacturing realities of today's marketplace. SPC can be effectively applied even with small lot sizes, complex parts, numerous part dimensions, and similar but different characteristics.

Take, for example, limitation 1: 15 to 25 data points are needed before calculating statistical control limits. To deal with this limitation, one should read chapter 15, entitled "*Control Charts for Small Lot Production Runs—The Short Run Charts.*" In that chapter, the reader will find three control charts developed specifically to deal with the problem of limited amounts of data and small part quantities.

For limitation 2, only one characteristic can be tracked per control chart, the reader should read chapter 19, "*Charts for Multiple Characteristics—The Group Charts.*"

To address limitation 3, even if part characteristics are similar, other differences such as material type or specification limits necessitate separate control charts, the reader should read chapter 11, entitled "*Control Charts for Similar Characteristics—The Target Charts.*"

But what about processes that have the combined problems of several part characteristics and limited data? Just read chapter 27, "*Charts for Short Runs and Multiple Characteristics—The Group Short Run Charts.*" Even combinations of the problems listed here will be addressed in this text.

This book should prove to be a help in identifying the best control chart(s) for almost any manufacturing situation. If the table of contents does not help identify the appropriate control chart, then the reader should refer to the control chart decision tree (Figure 6.1) in chapter 6. It is our hope that the reader finds the control chart decision tree (as well as this entire book) useful in solving the implementation problems that so many SPC practitioners face.

Finally, every chapter in this book contains examples that lead the reader through an examination of each SPC tool, including demonstrations of how the mathematics are performed. Readers should note that space limitations and an interest in keeping the text readable have required us to round the data contained in almost all calculations. In performing the math, we kept significant digits in intermediate calculation steps until final calculations were completed. It is hoped that following this convention will result in the highest accuracy in reported final values and estimates.

Chapter 2

SPC Basics

In today's computer-driven, technology-rich society, many companies find they have lots of data but little information. Data overload is the rule rather than the exception. In order to derive some understanding from the masses of data, one needs a tool—something that will help him or her make sense out of mountains of numbers. That tool is statistics. Basically, statistics is a powerful way of extracting meaningful information from numbers.

To illustrate how statistics can work in a real manufacturing environment, we will sample and analyze data from an autoclave. An autoclave is a large pressurized oven into which parts composed of layers of composite materials—such as fiberglass and plastic—held together by special adhesives are placed. This sandwich of materials is placed on a tool that, when heated and pressurized in an autoclave, will give form and shape to the composition of layers. After the process is complete, the output is a lightweight, super strong composite part.

Composite part quality is highly dependent upon autoclave temperatures. Of particular importance is the temperature of the autoclave during its cure cycle. It is important that the cure temperature fall within specifications of, say, 350°F ± 2 percent (343°F to 357°F). Temperatures must remain consistent not only within these stated specifications during the length of the curing cycle, but also from location to location on the parts within the autoclave. Thermocouples (T/Cs) are placed throughout the autoclave chamber and at various locations on the parts being cured. In this example, we will study only the readings of a single chamber T/C. To analyze the readings from all the T/Cs used during a cure would go beyond the purpose of this example.

Throughout the cure cycle, five temperature measurements are collected every 15 seconds with an automated data gathering device from one centrally located T/C. The 25 samples in Table 2.1 represent about 6.5 minutes of the cure cycle.

Once data have been gathered, it is natural to want to analyze them. For example, one may be interested in answering these questions.

1. What is the average autoclave temperature?
2. How much variation is there in the temperatures?

Table 2.1. Autoclave temperature measurements.

Sample number	X_1	X_2	X_3	X_4	X_5
1	351.17	348.57	348.57	350.92	353.90
2	350.26	355.26	358.44	346.87	353.33
3	344.31	350.79	349.73	351.54	349.28
4	359.64	347.81	347.29	352.89	355.41
5	355.63	350.51	351.07	346.46	353.09
6	354.40	349.08	348.73	345.32	352.11
7	353.10	357.19	349.07	344.29	349.84
8	349.95	343.96	346.28	341.83	349.23
9	350.39	351.92	344.22	349.57	346.58
10	345.96	351.87	345.74	347.59	352.91
11	348.13	345.04	346.55	347.77	356.79
12	345.77	342.16	349.70	346.89	351.83
13	355.34	356.75	360.22	351.06	352.56
14	348.50	347.07	344.00	351.50	346.69
15	342.98	343.53	347.80	352.90	352.81
16	350.09	352.42	347.26	348.91	348.06
17	350.73	352.80	345.78	347.59	353.91
18	347.91	356.31	344.30	348.34	352.24
19	351.91	353.94	342.94	347.71	355.85
20	347.02	351.49	354.36	352.74	357.86
21	349.61	347.75	351.35	343.02	351.91
22	350.15	345.22	348.44	346.13	353.72
23	352.32	342.85	352.18	352.50	353.72
24	347.43	353.61	348.66	352.44	353.51
25	351.29	349.31	355.12	342.21	352.75

3. Is the amount of variation in temperatures normal?
4. Do the temperatures change significantly during the curing cycle?
5. Do the temperatures fall within the engineering tolerances of 350°F ± 2 percent?

Reading the individual temperatures one-by-one in Table 2.1 would prove to be a painstaking and impractical way to answer these questions. One needs a straightforward way of summarizing the data. This is where the use of statistics can be an enormous help.

Statistics can be thought of as a collection of methods, procedures, and techniques that enable one to extract meaningful information from data. Some statistical tools allow one to place data into pictorial formats for summarization purposes. Take, for instance, a histogram (see Figure 2.1). A histogram is simply a graphical frequency display of data points in a bar chart format. The frequency of occurrence for any given data value is represented by the height of the bars. Histograms allow one to visualize a data set's variation, central values, and distributional shape.

Using a histogram allows one to more easily see information in a large amount of numbers. For example, notice that the temperature readings mostly cluster around the middle of the histogram and there are fewer data values at the edges. This phenomenon is characteristic of processes that follow a normal distribution.

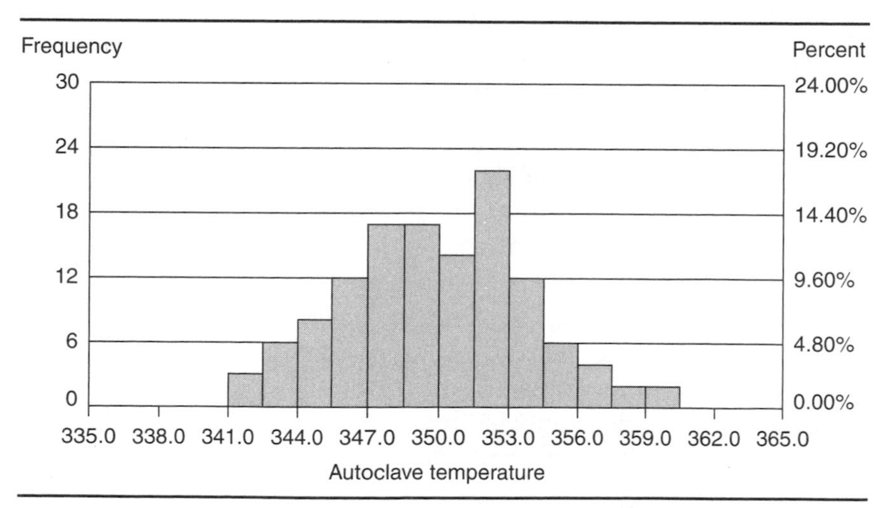

Figure 2.1. Autoclave temperature histogram.

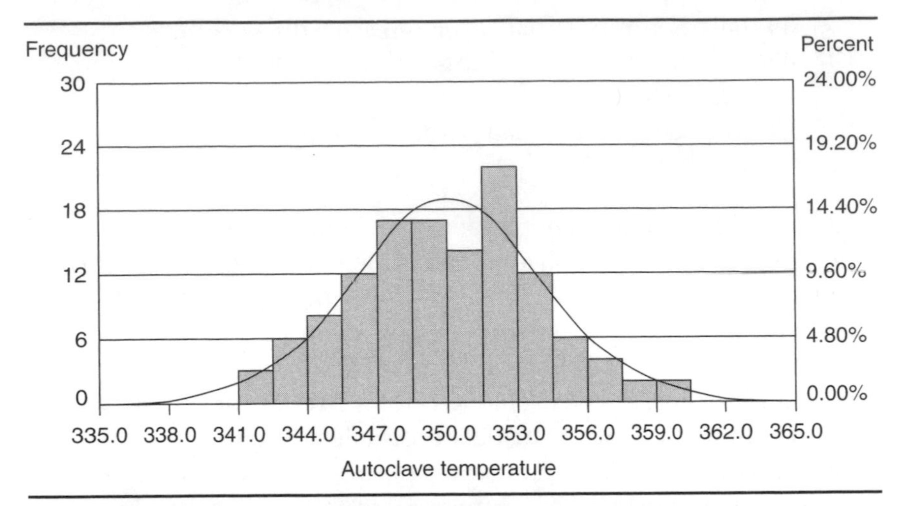

Figure 2.2. Autoclave temperature histogram with overlaid normal distribution curve.

The normal distribution curve can be overlaid on the histogram as seen in Figure 2.2.

Basically, the normal distribution is used to describe the behavior of data sets with the classic bell shape. It is a mathematical model that describes not only processes in nature (heights of human beings, for example), but also industrial processes (dimensional measurements, conductivity values, temperatures, and so on).

The normal distribution's location and spread are dependent, respectively, upon two population parameters.

1. μ (*mu*, the population mean)
2. σ (*sigma*, the population standard deviation)

μ and σ are both parameters that describe characteristics of some population. For example, if we wanted to calculate the true average height and standard deviation of 45 year olds in some country, we would have to measure the actual heights of all of that population's 45 year olds. This study would most likely be too costly and time-consuming for researchers to perform. Therefore, to get as close as possible to estimating the true values, researchers typically take only a small number of measures from the population of interest. These measures collectively form what is commonly called a *sample*. Then, the researchers estimate the true average and standard deviation with sample statistics.

To estimate μ, the population mean, one would calculate the arithmetic mean ($\overline{X}$, pronounced *X bar*) using Equation 2.1.

$$\overline{X} = \frac{\Sigma x_i}{n}$$

where $\overline{X}$ = arithmetic mean (average)
Σ = sum of
x_i = observed individual data values
n = number of individual data values in the sample

Equation 2.1. Formula for estimating the population mean.

To estimate σ, the population standard deviation, one would calculate the sample standard deviation (s) with Equation 2.2.

$$s = \sqrt{\frac{\Sigma(x_i - \overline{X})^2}{n - 1}}$$

where s = sample standard deviation
$\overline{X}$ = arithmetic mean (average)
Σ = sum of
x_i = observed individual data values
n = number of individual data values in the sample

Equation 2.2. Formula used for estimating the population standard deviation.

First, let's get clear on how to interpret these statistics. The average (or arithmetic mean) is a well-known and widely used statistic that describes a data set's central tendency. Measures of central tendency help one to identify the center or location of a data set.

Knowledge of the average is helpful, but measures of central tendency tell only part of the story. One must also consider the amount of dispersion or variation in the individual data values. Making decisions based only upon information about central tendency without information about variability can lead to erroneous conclusions.

For example, a person that drives 50 percent of the time in a highway's median and 50 percent of the time in its emergency lane is, on average, driving in the middle of the road. This conclusion, of course, is unsound because the analysis overlooks the obvious (and certainly dangerous) variability in the person's driving style. This is why we must also look at how much variability there is about the average.

Standard deviation is simply a measure of variability or spread in individual values. While the average describes the location of the normal curve, the standard deviation determines the curve's width (see Figure 2.3). The width of the normal curve increases with corresponding increases in the value of standard deviation. If the value of the standard deviation is small, then the curve width will be narrow. If the standard deviation value is large, then the width of the curve will be comparatively wide.

A convenient and useful aspect of the normal curve is that the area under the curve is known (see Figure 2.4). This is convenient in that, if we know the product specification and have good estimates of the population average and standard deviation, we can estimate the percentage of manufactured products that will fall within (or outside) the requirements. This can be helpful in determining how much scrap or rework can be expected from a given process and its specifications.

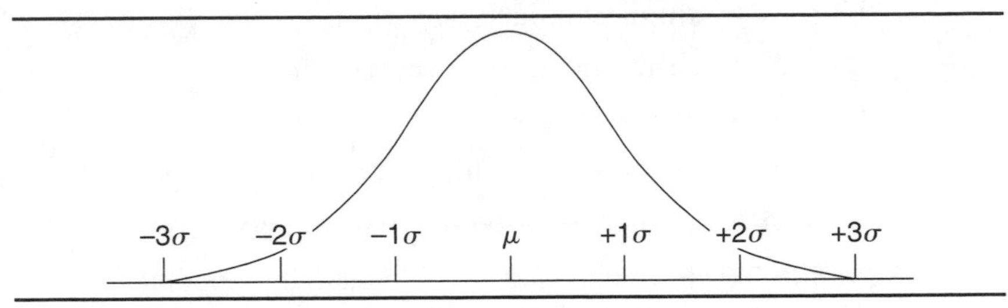

Figure 2.3. Illustration of a typical normal curve.

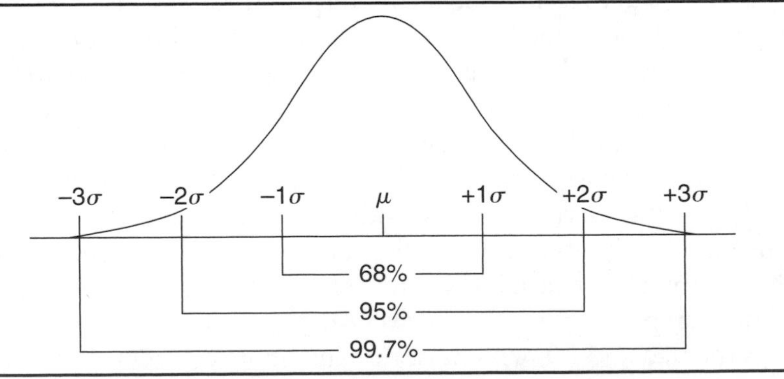

Figure 2.4. Areas of the normal curve.

Theoretically, the normal curve extends indefinitely on either side of the average. For practical purposes, essentially all data values (99.7 percent) fall between ± 3 sigma. Therefore, to make reasonable business decisions one needs to be primarily concerned with the six sigma spread of the bell curve.

All of this theory is useless unless it can be applied in a practical manner. Look again at the autoclave temperatures histogram in Figure 2.1. Visually, it seems as though the average temperature hovers around 350°F. The true process average can be estimated by summing all temperature readings and dividing by the number of readings. After applying Equation 2.1 to the autoclave data found in Table 2.1, the resulting overall average temperature is $\overline{\overline{X}} = 349.983$°F.

Once again, the average is useful to describe the center point of the data, but the variability must also be described. The standard deviation for all 125 temperatures is calculated using Equation 2.2. The result is $s = 3.98$°F.

With this information, we can now estimate the normal distribution for the autoclave temperatures (see Figure 2.5).

Given the ± 3 standard deviation spread of the data, Figure 2.5 illustrates that the autoclave temperatures are estimated to vary between 338.043°F and 361.923°F with an average of 349.983°F. This information can be compared with the specifications of 350°F ± 2 percent (see Figure 2.6). As Figure 2.6 illustrates, some T/C temperatures fall outside of the specification.

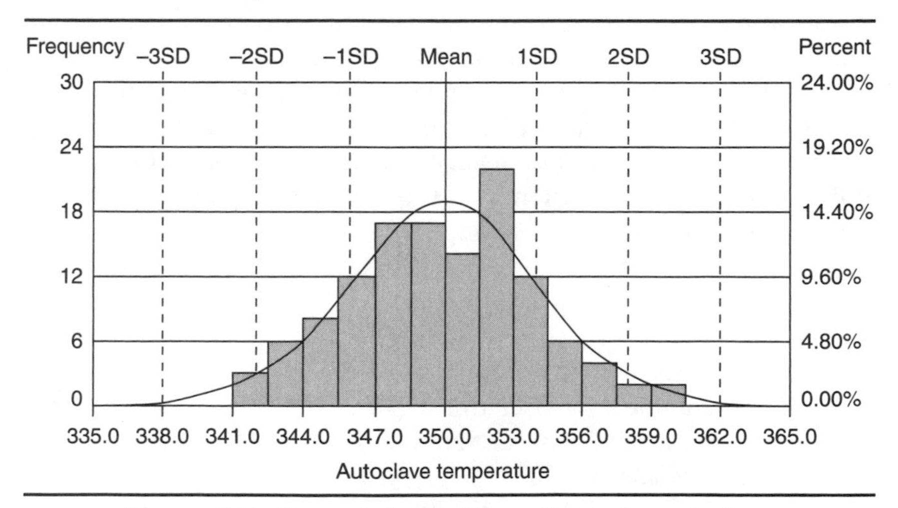

Figure 2.5. Normal distribution estimate for autoclave temperatures with standard deviation limits.

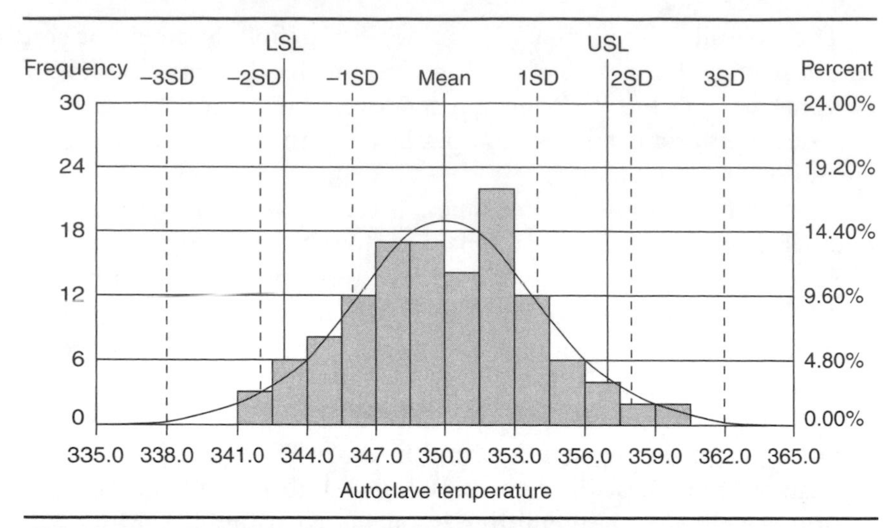

Figure 2.6. Relationship between the normal distribution of autoclave temperatures and their engineering specifications. (LSL = lower specification limit, USL = upper specification limit.)

Comparing the ±3 standard deviation process variation with engineering tolerances is equivalent to performing a process capability study. This specialized tool will be covered in chapter 4.

Recall the question earlier in the chapter "Do the temperatures change significantly during the curing cycle?" This question has not yet been answered and it cannot be answered with a histogram. The question can be answered only by carefully constructing and analyzing another statistical tool called a *control chart.*

Yes, histograms and normal curves are useful tools that allow one to visualize process location and variability. By themselves, though, neither histograms nor normal curves reveal any information considering time-to-time consistency. It is important to note that, before developing a histogram, sketching a normal curve, or performing a process capability study, one must first establish that the process is consistent over time. This can be done only with the use of a control chart.

Control charts are statistical tools used to determine if a process is changing significantly over time. They will reveal if a process is consistent or not. The basics of control charting are covered in the next chapter, "*Control Chart Overview.*"

Chapter 3

Control Chart Overview

In the last chapter, we looked at a histogram to evaluate a data set's average and standard deviation. By using a histogram we were able to visualize a large amount of data all at once. However, no matter how useful histograms are, they do not reflect the element of time. Because of this lack of time order, any information gained solely from a histogram is suspect. The reason is that, over time, any change in the average or standard deviation is not revealed by histograms. To be alerted to these changes, one needs a time-ordered tool called a *control chart.*

Basically, control charts are tools used to determine if a process is in control or out of control. In statistical language, a process that is in control is one whose average and standard deviation are known and predictable. Conversely, an out-of-control process is one whose either average or standard deviation is changing, thus unpredictable.

- *In control*—Stable, predictable, consistent, unchanging
- *Out of control*—Unstable, unpredictable, inconsistent, changing

To determine process stability, one needs a control chart. All control charts have three common elements.

1. Plot points. These typically represent individual measurements, averages, standard deviations, or ranges.

2. A centerline. Typically, though not always, the centerline is the average of the points plotted on the chart.
3. Control limits.

A typical looking control chart is illustrated in Figure 3.1. The data points and centerline are self-explanatory. What may need explanation, though, are the lines called *control limits.* Control limits define the amount of expected variability in a series of measurements. Variation is present in all things; the challenge is to characterize what is and is not natural variation.

Control limits represent boundaries of variability that indicate 99.7 percent of a measured characteristic's natural variation (see Figure 2.3). Control limits are typically set at ±3 standard deviations from the average. For variables data, two control charts are used to describe the variation of a measured characteristic: one chart to evaluate the stability of the process average and another to describe the stability of the variation of individual data values.

Refer to the autoclave temperatures in Table 2.1. If one were to arrange the data points into subgroups of five, an average and standard deviation could be calculated for each subgroup. Conveniently, each subgroup's average and standard deviation values can then be plotted and evaluated with respect to time.

Calculating the average is easy enough, but calculating each subgroup's standard deviation is more complex (see Equation 2.2). So, instead of calculating the rather intricate standard deviation statistic, one could compute a statistic called the *range* (R). Like the standard deviation, the range is a commonly used statistic for describing a subgroup's variability. The range is calculated by subtracting the smallest measurement in a subgroup from the largest.

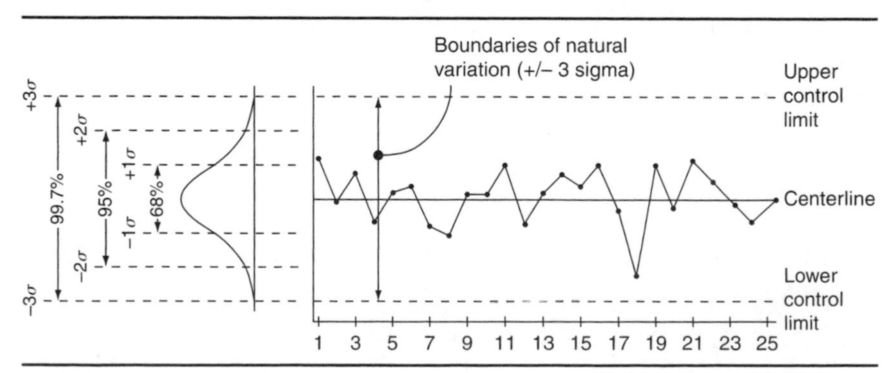

Figure 3.1. A typical looking control chart with normal distribution.

Table 3.1. Autoclave temperature measurements with $\overline{X}$ and range statistics calculated for each subgroup.

Subgroup	Temperatures					$\overline{X}$	Range
1	351.17	348.57	348.57	350.92	353.90	350.626	5.33
2	350.26	355.26	358.44	346.87	353.33	352.832	11.57
3	344.31	350.79	349.73	351.54	349.28	349.130	7.23
4	359.64	347.81	347.29	352.89	355.41	352.608	12.35
5	355.63	350.51	351.07	346.46	353.09	351.352	9.17
6	354.40	349.08	348.73	345.32	352.11	349.928	9.08
7	353.10	357.19	349.07	344.29	349.84	350.698	12.90
8	349.95	343.96	346.28	341.83	349.23	346.250	8.12
9	350.39	351.92	344.22	349.57	346.58	348.536	7.70
10	345.96	351.87	345.74	347.59	352.91	348.814	7.17
11	348.13	345.04	346.55	347.77	356.79	348.856	11.75
12	345.77	342.16	349.70	346.89	351.83	347.270	9.67
13	355.34	356.75	360.22	351.06	352.56	355.186	9.16
14	348.50	347.07	344.00	351.50	346.69	347.552	7.50
15	342.98	343.53	347.80	352.90	352.81	348.004	9.92
16	350.09	352.42	347.26	348.91	348.06	349.348	5.16
17	350.73	352.80	345.78	347.59	353.91	350.162	8.13
18	347.91	356.31	344.30	348.34	352.24	349.820	12.01
19	351.91	353.94	342.94	347.71	355.85	350.470	12.91
20	347.02	351.49	354.36	352.74	357.86	352.694	10.84
21	349.61	347.75	351.35	343.02	351.91	348.728	8.89
22	350.15	345.22	348.44	346.13	353.72	348.732	8.50
23	352.32	342.85	352.18	352.50	353.72	350.714	10.87
24	347.43	353.61	348.66	352.44	353.51	351.130	6.18
25	351.29	349.31	355.12	342.21	352.75	350.136	12.91

The data in Table 3.1 are the autoclave temperatures from the previous chapter. The two right columns in the table display the average and range values for each subgroup. These statistics will be used to construct variables data control charts called *$\overline{X}$ and range charts.*

The average and range values from Table 3.1 have been plotted on the $\overline{X}$ and range chart in Figure 3.2. These charts allow one to visualize variability of both the average ($\overline{X}$) and the range values through time.

Recall that control limits define the amount of expected variability for subgroup averages and ranges (the plotted points in Figure 3.2). One can consider the control limits to be estimates of natural process variation when the process is in control. Control chart rules used to signal an out-of-control process will be covered later in this chapter.

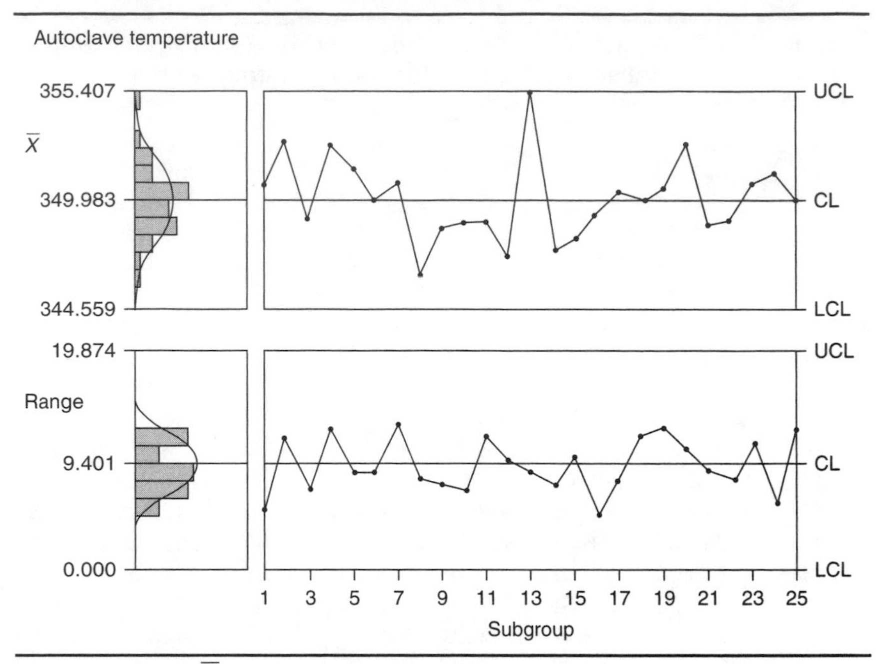

Figure 3.2. $\overline{X}$ and range chart of autoclave temperature readings.

In interpreting and evaluating control charts, one should always start with the range chart. For the range chart calculations, see Calculation 3.1. The overall average range ($\overline{R}$) for all 25 subgroups is 9.401°F. This value will be used in control limit calculations for both the $\overline{X}$ and the range chart.

$$\overline{R} = \frac{\sum R}{k} = \frac{235.02}{25} = 9.401$$

$$UCL_R = D_4\overline{R} = 2.114(9.401) = 19.874$$

$$LCL_R = D_3\overline{R} = 0(9.401) = 0$$

Calculation 3.1. Control chart calculations for the autoclave temperature range chart. (See Table 3.2 for D_4 and D_3 values.)

The upper control limit for the range chart (UCL_R) is 19.874°F. This means that the range value for any subgroup could naturally be expected to be as large as 19.874°F. If any range value were larger than the UCL_R, this would indicate a statistically significant increase in process variability. An out-of-control situation merits an examination of the process to determine the specific cause for the process change.

Table 3.2. Table of control chart constants (a portion of Table A.1 in the appendix).

n	A_2	D_3	D_4
1	2.660	0	3.267
2	1.880	0	3.267
3	1.023	0	2.574
4	0.729	0	2.282
5	0.577	0	2.114
6	0.483	0	2.004
7	0.419	0.076	1.924
8	0.373	0.136	1.864
9	0.337	0.184	1.816
10	0.308	0.223	1.777

The lower control limit for this range chart (LCL_R) is zero since D_3 values are zero for subgroup sizes (n) from 1 to 6 (see Table 3.2). Looking again at the range chart in Figure 3.2, notice that all of the range values fall within the natural bounds defined by the control limits. In essence, this control chart indicates that the standard deviation in autoclave temperatures (expressed here as range values) is stable.

Once the range chart has been analyzed, one should next evaluate the $\overline{X}$ chart. Notice that, like the range values, all of the $\overline{X}$ values fall within the natural boundaries defined by its control limits. The $\overline{X}$ value of 355.186°F for subgroup number 13 does, in fact, fall within the $UCL_{\overline{X}}$ of 355.407°F. Therefore, this is not considered an out-of-control situation. This $\overline{X}$ chart shows that the overall average autoclave temperature $\overline{\overline{X}}$ is 349.983°F (the $\overline{X}$ chart centerline.) Because this process is in control, $\overline{\overline{X}}$ can be considered to be a reliable estimate of the true process average.

$$\overline{\overline{X}} = \frac{\Sigma \overline{X}}{k} = \frac{8749.576}{25} = 349.983$$

$$UCL_{\overline{X}} = \overline{\overline{X}} + A_2\overline{R} = 349.983 + 0.577(9.401) = 355.407$$

$$LCL_{\overline{X}} = \overline{\overline{X}} - A_2\overline{R} = 349.983 - 0.577(9.401) = 344.559$$

Calculation 3.2. Control chart calculations for the autoclave temperature $\overline{X}$ chart. (See Table 3.2 for values of A_2.)

The chart also shows that the $\overline{X}$ plot points should typically vary between 355.407°F and 344.559°F, the UCL and LCL respectively for the $\overline{X}$ chart. The formulas used for calculating the $\overline{X}$ control chart limits are found in Calculation 3.2.

In summary, the autoclave temperatures are in control. We can now say, for the time period shown, the average and standard deviation of the autoclave temperatures are unchanging, consistent, stable, and predictable. It is like having one normal curve that repeats itself again and again over time (see Figure 3.3).

When a process is in control, the following benefits can be realized.

- Percent of product fallout can be accurately predicted.
- Scrap and rework estimates can be predicted prior to running a job.
- Machine settings can be adjusted to optimize throughput.
- Engineers can incorporate statistical tolerancing into their drawings. This can mean increased component tolerances without compromising assembly performance.
- Product designs can be statistically modeled to accurately predict fit and performance yields prior to prototype assembly.
- Machine utilization can be optimized. For example, high-precision machines and resources will not be wasted on manufacturing low-precision dimensions.
- Process improvement resources will be better spent.

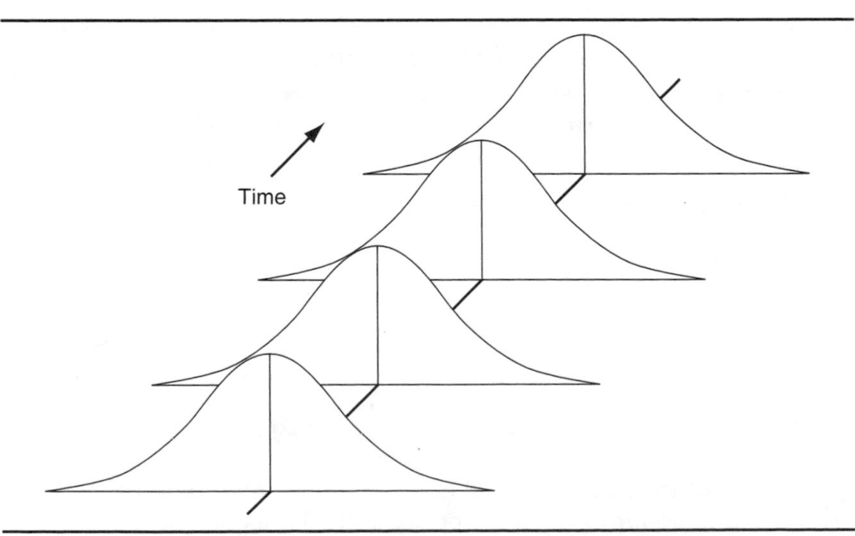

Figure 3.3. Illustration of an in-control process.

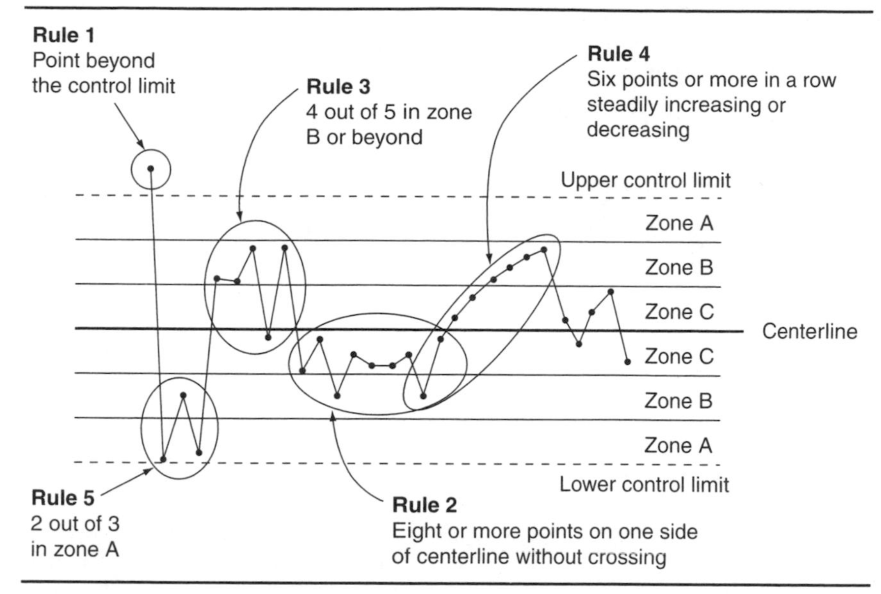

Figure 3.4. Assignable cause patterns on a control chart. (Graphic obtained from the Boeing Company's D1-9000® document, *Advanced Quality System® for Boeing Suppliers*. Reproduced with permission.)

Many benefits can be gained from in-control processes. However, not all processes show stability and predictability. Generally, there are five control chart rules (shown graphically in Figure 3.4) that are commonly used to signal a change in either the process average or variation.

- Rule 1: Points beyond the control limits
- Rule 2: Eight or more consecutive points either above or below the centerline
- Rule 3: Four out of five consecutive points in or beyond the 2 sigma zone (referred to in Figure 3.4 as *zone B*)
- Rule 4: Six points or more in a row steadily increasing or decreasing
- Rule 5: Two out of three consecutive points in the 3 sigma region (referred to as *zone A* in Figure 3.4)

Figures 3.5 through Figure 3.7 are examples of control charts that signal changes in either average or standard deviation (expressed using ranges in these examples). Notice the relationship between the histogram patterns and their respective control charts. When a bell curve shifts to the right, the plot points on the $\overline{X}$ chart go up. A shift to the left, the plot points go down. When a bell curve widens,

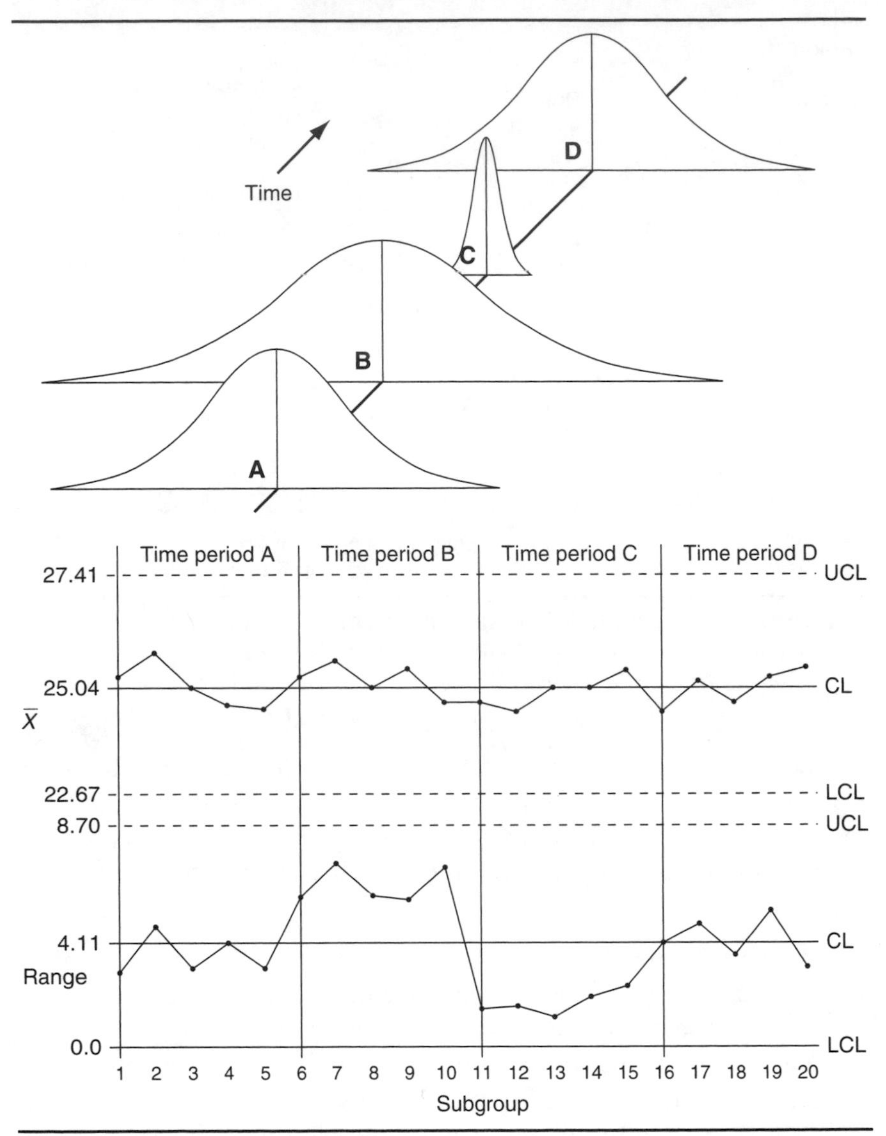

Figure 3.5. A process that is out of control with respect to standard deviation. Control limits on the $\overline{X}$ chart are unreliable.

the plot points on the range chart go up. When the curve narrows, the plot points go down.

Control charts graphically illustrate process changes. Using the $\overline{X}$ and range chart combination is especially beneficial because they can independently differentiate between changes in the process average and changes in the process standard deviation.

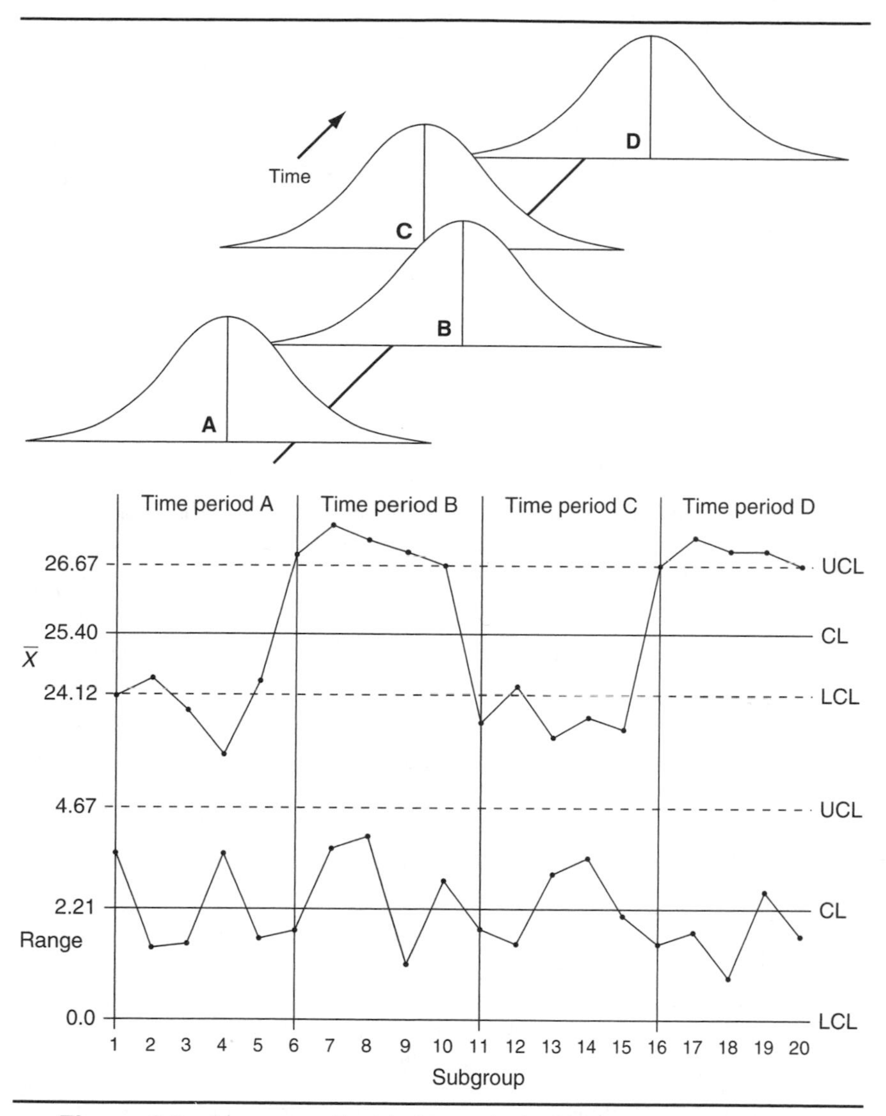

Figure 3.6. A process that is in control with respect to standard deviation, but out of control with respect to average.

Note that the control limits for the $\overline{X}$ charts in Figures 3.5 and 3.7 are unreliable because the range charts in each figure are out of control.

If a range chart is out of control, its $\overline{R}$ value is unreliable. Notice in Calculation 3.2 that the $\overline{R}$ is used in calculating control limits for

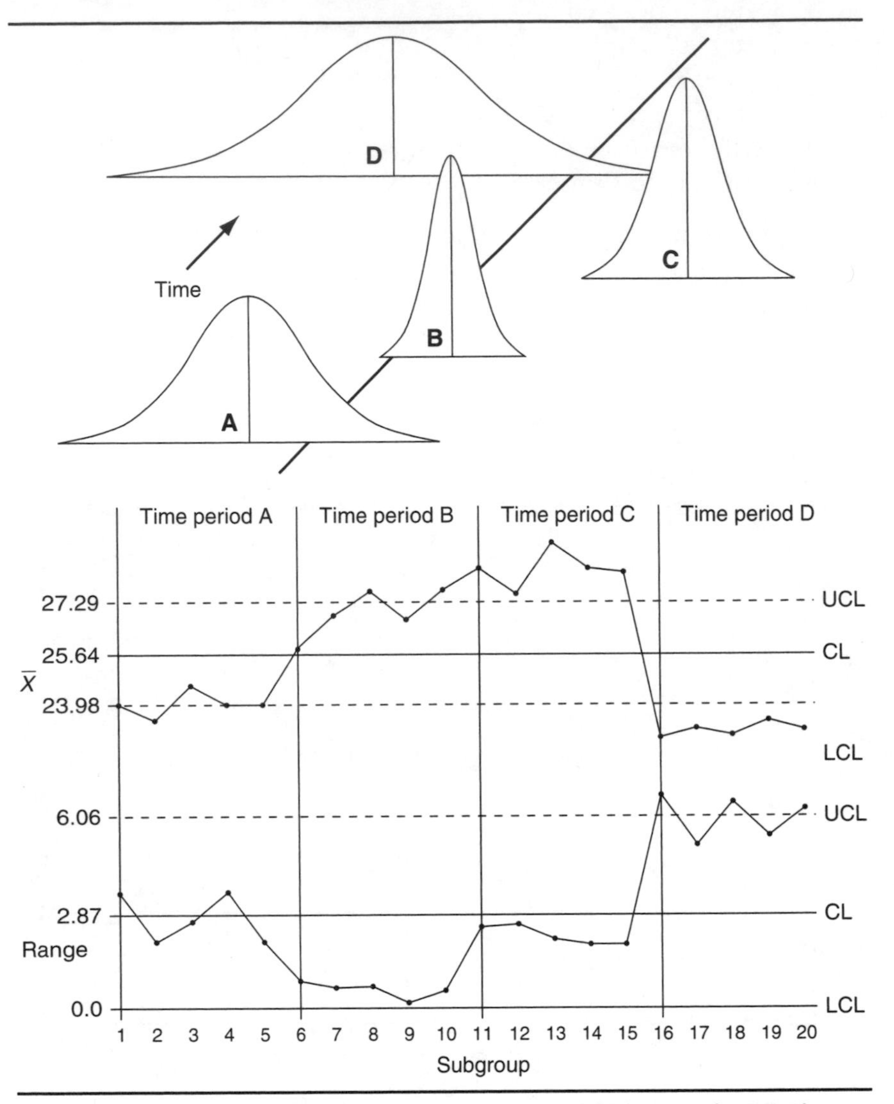

Figure 3.7. Statistical chaos. A process that is out of control with respect to standard deviation and out of control with respect to average.

the $\overline{X}$ chart. Because the $\overline{R}$ value is unreliable, so will be the $\overline{X}$ chart control limits.

At first, it may seem as though control charts with assignable causes are a problem. But out-of-control charts actually reveal much useful information about processes. When a process is out of control, one may be able to

- Detect unwanted process changes.
- Detect desirable process changes.
- Prove that a process change was or was not an improvement.
- Determine when to make a change to the process.
- Verify measurement system improvements.
- Verify process improvements.

Regardless of a process' predictability, or lack thereof, many advantages exist by employing the use of control charts. Notice that, if control charts are not used, the benefits just outlined simply vanish.

So far, the discussion in this chapter has centered solely on identifying a process as in control or out of control. While determining process control is vitally important, we have yet to evaluate the process' ability to conform to engineering specifications or customer requirements. This ability is critical because processes exist to serve customers. To compare actual process behavior with specifications, one can use tools called *process capability* and *performance studies,* which will be covered in the next chapter.

Chapter 4

Process Capability and Performance Studies

In the previous chapter, control charts were introduced to determine if a process is in control over some period of time. While issues of process control address stability and consistency, control says nothing about acceptability. It is possible to have a stable, consistent process that produces 100 percent unacceptable output. The challenge, then, is not only to evaluate a process's stability, but also its capability and its acceptability. If a process is in control, then its behavior can be reliably compared to engineering tolerances. The two methods used to make these comparisons are called *process capability* and *performance studies.*[1]

Let's first look at process capability studies. Process capability studies are used to compare the natural variation of individual data values to engineering specifications (tolerances). Like its name, process capability studies only indicate what the process is capable of producing, not what it is actually producing.

Prior to performing a process capability study, one must ensure that the following requirements are met.

1. The process variability is stable (as indicated by an in-control R chart or s chart.)
2. The histogram of individual data values is approximately normal.

3. The engineering tolerances are known.
4. The process standard deviation estimate is known.

The first two requirements have already been met for the autoclave example discussed in chapters 2 and 3. In addition, the engineering tolerance for the autoclave temperatures has been previously identified as 350°F ± 2 percent (350°F ± 7°F). However, we still need the value for the process standard deviation.

Recall that calculations for the *sample* standard deviation s were covered in chapter 2. However, we have yet to estimate the process standard deviation, $\hat{\sigma}$ (pronounced *sigma hat*). (In statistical language, the hat above σ means *estimate of*. So, $\hat{\sigma}$ is statistical shorthand for *the estimate of the process standard deviation*.)

In the autoclave example the variability of the process was consistent with an overall average range $\overline{R}$ of 9.401°F. Because the range chart was consistent, one can estimate the value of the process standard deviation, $\hat{\sigma}$, by using both the centerline ($\overline{R}$) from the range chart and d_2, a statistical constant. The formula to convert $\overline{R}$ into $\hat{\sigma}$ is shown in Equation 4.1.

$$\hat{\sigma} = \frac{\overline{R}}{d_2}$$

Equation 4.1. Formula for estimating σ.

The constant d_2 found in Equation 4.1 is based upon subgroup size (n). In our case $n = 5$, therefore, $d_2 = 2.326$ (see Table A.1 in Appendix A).

The estimated standard deviation for the autoclave temperature is found in Calculation 4.1.

$$\hat{\sigma} = \frac{\overline{R}}{d_2} = \frac{9.401°}{2.326} = 4.04°$$

Calculation 4.1. Estimation of autoclave temperature standard deviation.

Process capability is a comparison between the natural variation in some process characteristic and its specifications. Knowing the value of $\hat{\sigma}$ will help one to figure out how much natural variation occurs in the process. The natural amount of process variation (6 sigma) can be interpreted as the range over which almost all individual measurements of a population can be expected to fall (see Figure 4.1).

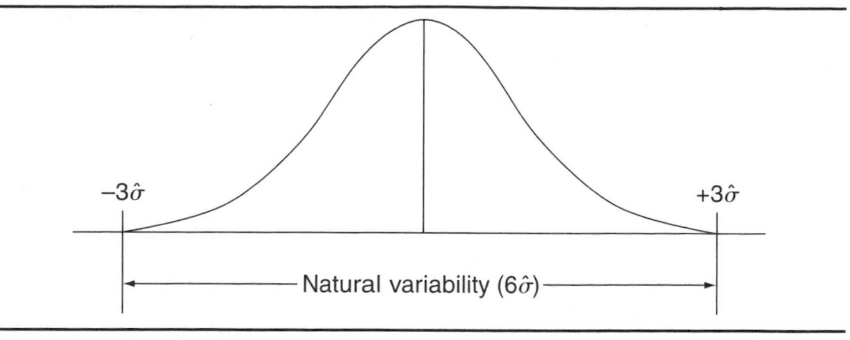

Figure 4.1. Natural variation of a process, estimated as $6\hat{\sigma}$.

Performing a Process Capability Study

Performing a process capability study means comparing a process' natural variability against engineering tolerances to determine if the variability is capable of "fitting" within the requirements (see Figure 4.2). When a process is not centered on a target, process capability studies do not always indicate the likelihood of producing nonconforming products, but they are a good gauge for how a process might perform.

C_p Ratio

To numerically quantify the relationship between natural variability and engineering tolerance, one can calculate a C_p (process capability)

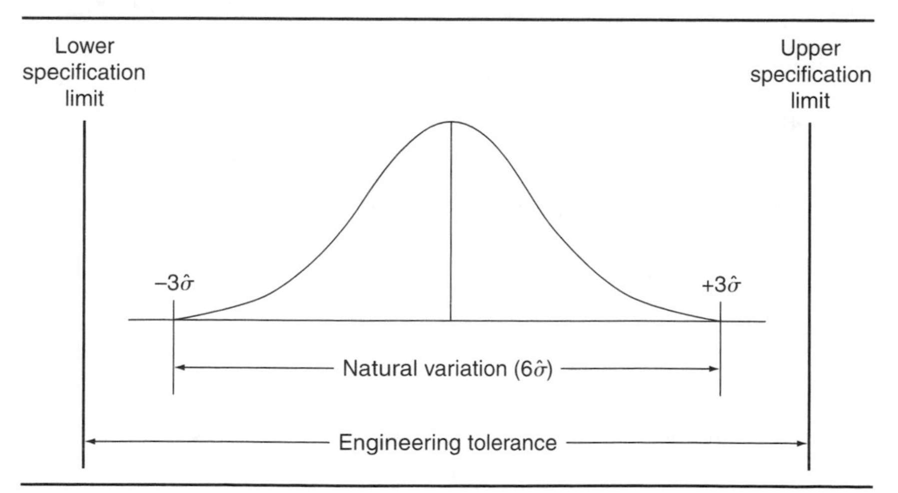

Figure 4.2. Comparison of the width of a normal curve to engineering tolerance; a process capability study.

ratio. The C_p ratio is a mathematical way to express Figure 4.2's pictorial relationship between a process' engineering tolerance and its natural variation. C_p is calculated by dividing the total engineering tolerance by the $6\hat{\sigma}$ spread of the normal curve (see Equation 4.2).

$$C_p = \frac{\text{Upper specification limit (USL)} - \text{Lower specification limit (LSL)}}{6\hat{\sigma}}$$

Equation 4.2. C_p ratio equation.

Because of the nature of this calculation, it is desirable that the value of C_p be greater than 1. If C_p is greater than 1 it means that the natural variation is less than the engineering tolerance. In effect, a C_p greater than 1 indicates the process is capable of producing nearly 100 percent acceptable output. Figure 4.3 displays some general guidelines for interpreting the C_p index.

The process capability index is a powerful and useful communication tool. Given a single C_p value, one immediately knows how a process' natural variation compares to its engineering tolerances. This one C_p value also communicates information about the quality of a process. That is, C_p can be used to indicate how much product might be expected to fall outside engineering tolerances (see Table 4.1).

Even though C_p values can be enlightening, they do not take into consideration the stability of the process average nor whether the overall process average is centered. A process is centered when the overall process average falls directly in the middle of two-sided engineering tolerances. Because the C_p index fails to address process centering or the lack thereof, the C_{pk} process performance ratio was developed.

Process Performance and the C_{pk} Ratio

Imagine that a process has very little variation when compared with its engineering tolerances. This is a good thing. It is good because the C_p ratio will be greater than 1, indicating a more than capable process. However, to the uninitiated, a C_p greater than 1 could be misleading. Imagine also that the same process has an overall average that is far outside the USL and that no part of the estimate of the normal curve falls within the engineering tolerances.

One might assume that, because the C_p value is greater than 1, it means that the process is currently producing almost 100 percent ac-

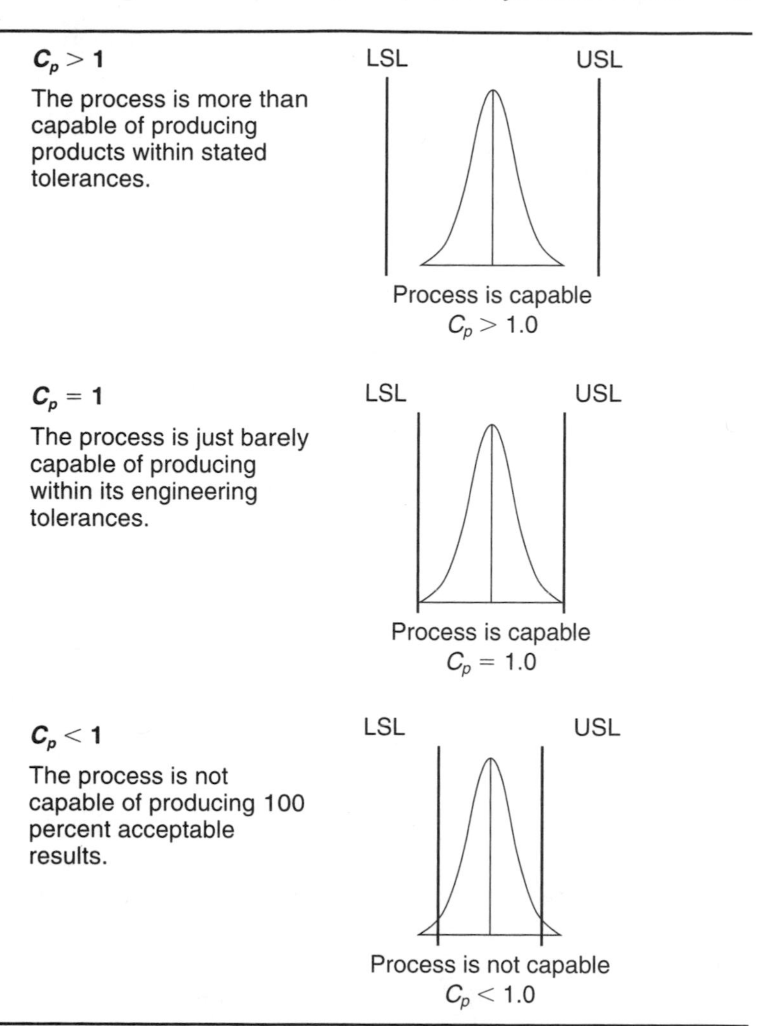

Figure 4.3. Interpreting the C_p index.

ceptable output. This is a false assumption. Recall that the C_p ratio indicates only what a process is capable of producing. In this case, the process is capable, but not producing any good product. This is what the C_p ratio has been roundly criticized for—it indicates nothing about process performance. Instead, it indicates only what a process is capable of doing.

Unlike C_p values, C_{pk} ratios take into consideration the location of the overall average and, therefore, give a better indication of how the process is performing relative to upper and lower specification limits.

Table 4.1. Process fallout based on C_p values.

Calculated C_p	Predicted fallout (if process is centered and stable)
0.5	133,620 ppm
0.6	71,860
0.7	35,730
0.8	16,396
0.9	6,934
1.0	2,700
1.1	966
1.2	318
1.3	96
1.4	26
1.5	7
1.6	2
1.7	340 ppb
1.8	60
1.9	12
2.0	2

ppm = parts per million
ppb = parts per billion
(10,000 ppb = 1%)

C_{pk} is a unitless ratio that compares process statistics to engineering tolerances. Specifically, C_{pk} is a performance ratio that takes into account a process' variability and centering and compares them explicitly to each of the upper and lower specifications. Also, the C_{pk} ratio focuses on the worst case scenario. That is, the reported C_{pk} value reflects the specification limit that resides closest to the process average.

To ensure that the C_{pk} values reported from a process performance study are reliable, one must ensure that the following requirements are met.

1. Both the process variability and average are stable (as indicated by an in-control X and MR chart, $\overline{X}$ and range chart, or $\overline{X}$ and s chart).
2. The histogram of individual data values is approximately normal.
3. The engineering tolerances are known.
4. Reliable estimates of the process average and standard deviation are known.

Assuming that these requirements have been met, a reliable C_{pk} value can be calculated. Note that two calculations must be performed in determining the value of C_{pk}. The C_{pk} value that should be reported is the smaller number resulting from the two calculations performed in Equation 4.3. The C_{pk} formula is illustrated in Figure 4.4.

$$C_{pk\ upper} = \frac{USL - \bar{\bar{X}}}{3\hat{\sigma}}$$

$$C_{pk\ lower} = \frac{\bar{\bar{X}} - LSL}{3\hat{\sigma}}$$

Equation 4.3. Formulas for C_{pk} calculation. The reported C_{pk} is the smaller number resulting from the two calculations.

Here are some general guidelines for interpreting C_{pk}.

1. If $C_{pk} > 1$, then the process is more than capable of producing products within stated tolerances.
2. If $C_{pk} = 1$, then the process is just barely capable of producing products within its engineering tolerances.
3. If $C_{pk} < 1$, then the process is not capable of producing 100 percent acceptable products.
4. If C_{pk} has a negative value, then the overall process average is outside of one of the specification limits.

Table 4.2 can be used to indicate C_{pk} fallout based on a centered process.

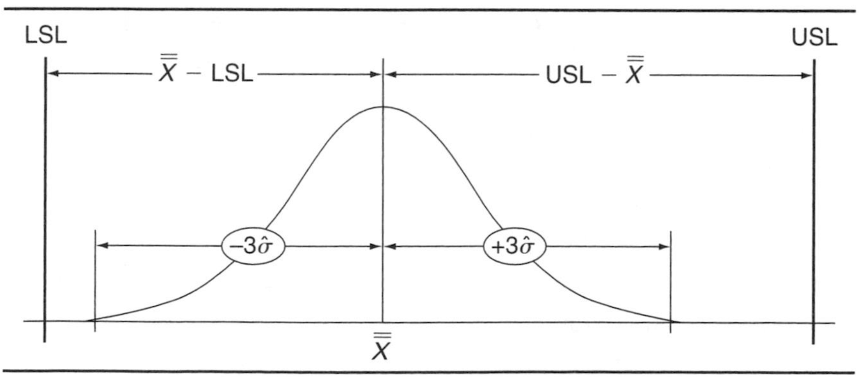

Figure 4.4. Illustration of C_{pk} formula.

Table 4.2. C_{pk} fallout based on a centered process.

Calculated C_{pk} ratio	Predicted fallout (if process is centered)	Predicted fallout in each tail of bell curve (if process is centered)
0.5	133,620 ppm	66,810 ppm
0.6	71,860	35,930
0.7	35,730	17,865
0.8	16,396	8,198
0.9	6,934	3,467
1.0	2,700	1,350
1.1	966	483
1.2	318	159
1.3	96	48
1.4	26	13
1.5	7	3
1.6	2	1
1.7	340 ppb	170 ppb
1.8	60	30
1.9	12	6
2.0	2	1

ppm = parts per million
ppb = parts per billion
(10,000 ppb = 1%)

Relationship Between C_p and C_{pk}

C_p and C_{pk} are best used together to evaluate process performance and acceptability. Here are some general rules when evaluating C_p and C_{pk} together.

1. C_{pk} can be equal to but never larger than $C_{p.}$
2. C_p and C_{pk} are equal only when the process is centered.
3. If C_p is larger than $C_{pk,}$ then the process is not centered.
4. If both C_p and C_{pk} are greater than 1, the process is capable and performing within tolerances.
5. If both C_p and C_{pk} are less than 1, the process is not capable and not performing within tolerances.
6. If C_p is greater than 1 and C_{pk} is less than one, the process is capable, not centered, and not performing within specifications.

The visual relationship between C_p and C_{pk} is illustrated in Figure 4.5.

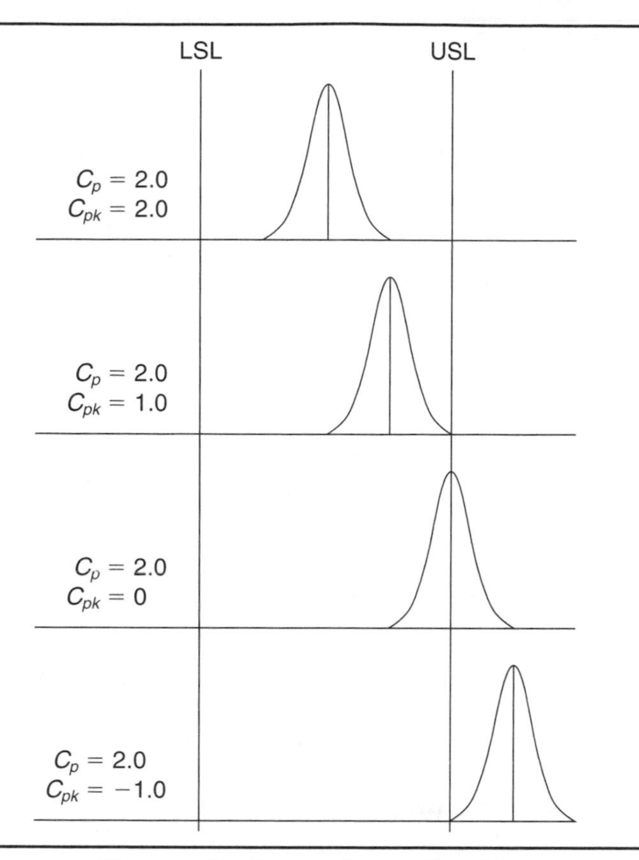

Figure 4.5. C_p and C_{pk} relationships.

Steps to Complete Process Capability and Performance Studies

There are four steps necessary to perform process capability and performance studies. The autoclave temperatures example will be used to illustrate the steps.

Step 1. Draw an estimate of the normal curve of the process. This step allows one to construct a picture of how much natural variation occurs in the individual measurements (see Figure 4.6).

Step 2. Label the bell curve with the $\overline{\overline{X}}$ and ±3 standard deviation limits (see Calculation 4.2).

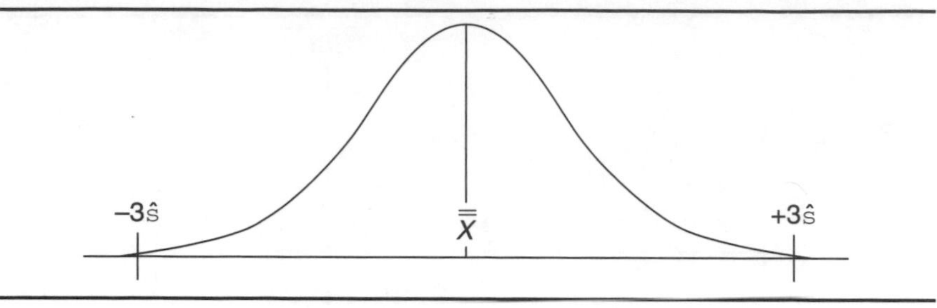

Figure 4.6. The normal curve.

$$\bar{\bar{X}} \pm 3\hat{\sigma}$$
$$= 349.983° \pm 3(4.04°)$$
$$= 349.983° \pm 12.12°$$
$$\{337.863°, 362.103°\}$$

Calculation 4.2. Estimate of the boundaries of natural variation.

Figure 4.7 illustrates that the highest estimated temperature during the autoclave cure cycle for the T/C being evaluated is 362.103°F. The lowest estimated temperature is 337.863°F.

Step 3. Draw vertical lines on the bell curve to represent the USL and LSL. The specifications are 350°F ± 7°F (see Figure 4.8).

Step 4. Calculate the C_p and C_{pk} ratios. (See Calculations 4.3 and 4.4).

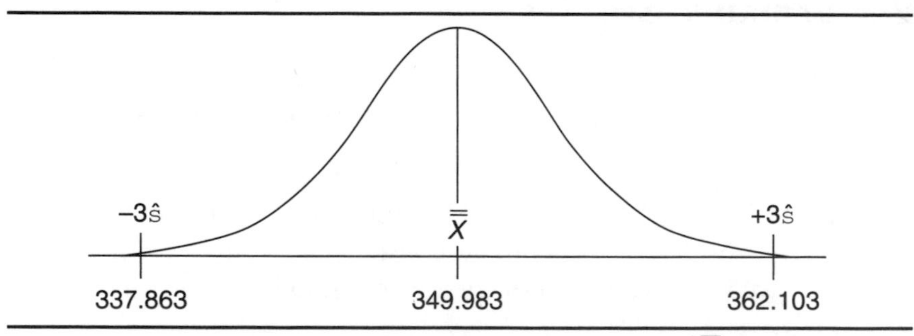

Figure 4.7. Normal curve labelled with the $\bar{\bar{X}}$ and ±3 standard deviation limits.

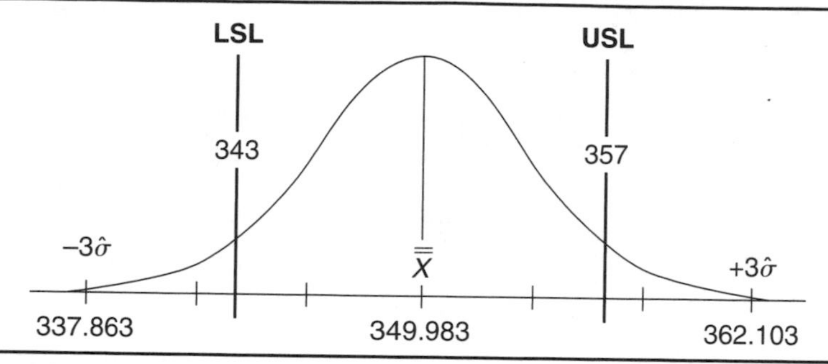

Figure 4.8. Estimate of the normal curve for the autoclave temperatures with engineering tolerances.

The C_p ratio is

$$
\begin{aligned}
C_p &= \frac{\text{USL} - \text{LSL}}{6\hat{\sigma}} \\
&= \frac{357^\circ - 343^\circ}{6(4.04^\circ)} \\
&= \frac{14}{24.24} \\
&= 0.58
\end{aligned}
$$

Calculation 4.3. C_p calculations for autoclave temperatures.

The C_p value of 0.58 is less than 1, therefore the process is not capable of producing 100 percent acceptable product. Based on Table 4.2, if the process were centered exactly on 350°F, the expected potential fallout is about 71,860 ppm (7.2 percent).

The C_{pk} ratio is

$$
\begin{aligned}
C_{pk\ upper} &= \frac{\text{USL} - \bar{\bar{X}}}{3\hat{\sigma}} \\
&= \frac{357^\circ - 349.983^\circ}{3(4.04^\circ)} \\
&= \frac{7.017}{12.12} \\
&= 0.58
\end{aligned}
$$

$$
\begin{aligned}
C_{pk\ lower} &= \frac{\overline{\overline{X}} - \text{LSL}}{3\hat{\sigma}} \\
&= \frac{349.983° - 343°}{3(4.04°)} \\
&= \frac{6.983}{12.12} \\
&= 0.58
\end{aligned}
$$

Calculation 4.4. C_{pk} calculations for autoclave temperatures.

Both C_{pk} values are 0.58. Therefore, the reported C_{pk} ratio is 0.58. Because the C_p is also 0.58, this means that the process is centered. (The process is not exactly centered on 350°F, but the $\overline{\overline{X}}$ of 349.983°F is close enough to 350°F to result in the same C_{pk} values for each of the two calculations.)

With the ratios discussed in this chapter, one has powerful ways of communicating process capability and performance. With only two numbers (C_p and C_{pk}) one can communicate or learn a great deal about a process and its ability to conform to engineering requirements.

Note

1. If a histogram reveals a nonnormal process distribution, advanced techniques not discussed in this text are required to complete the capability study. Also, autocorrelated and within-piece variability are phenomena not discussed in this text.

Chapter 5

Data Collection Strategies

Once the decision has been made to use statistical tools, confusion is typically found around how data should be gathered. Therefore, prior to the use of any statistical tool, one must develop a coherent data collection plan. This plan is called a *data collection strategy*.

Data collection strategies greatly influence interpretation of control charts and process capability studies. That is, how and when measurements are taken may bias conclusions made from control charts. Generally, though, the determination of a data collection strategy is very subjective. This chapter will present many solutions to the challenges of sampling, subgrouping, and collecting data.

However, we want to note that sampling and subgrouping is a complex topic and that a thorough and complete examination of this subject is beyond the scope of this text. Instead, we want to cover information that will address most of the challenges posed by the subject of sampling and subgrouping.

Prior to developing a data collection strategy, questions such as these should be asked and answered.

- Should measurements be made?
- What should be measured?
- When should data be taken?
- How often should data be taken?

- Should the data be subgrouped and, if so, what should be the subgroup size?

Should Measurements Be Made?

Data collection costs money. Data should be collected when there is reasonable belief that doing so will benefit the organization. The information gained from the data should outweigh the money spent on collecting it. Only a small amount of data is initially required to confirm whether or not further data collection will be beneficial. There are four basic reasons for collecting data.

1. To comply with customer or industry requirements.
2. To learn an unknown. For example, an engineer may have a predicted estimate of a unit's performance based on preproduction calculations, however, actual data are necessary to confirm the engineer's estimates.
3. To optimize. Exploratory measurements in the form of designed experiments, correlation studies, and regression analysis are all methods used to find optimal levels of process parameters.
4. For process control.

Ongoing data collection should be pursued only if it is a customer requirement or it continues to provide benefits to the organization.

Determining What to Measure

Before knowing what to measure, one must decide what needs to be learned. Answering the question, "What am I trying to learn?" should provide clarity to the question, "What do I measure?"

If fit problems are occurring during assembly, then data from mating surfaces should be collected. If chemical depletion in an anodize bath needs to be predicted, one should base the data collection on the surface area being processed.

Once again, collecting data costs money. When process operators are collecting data, parts are either not being made or attention is being diverted away from manufacturing duties. Either way, there is a real or perceived cost of collecting data. These costs should be compared against the potential benefit (or detriment) of collecting the data (or not collecting it). In other words, use common sense when establishing data collection strategies.

When to Collect Data

Data should be collected at the earliest possible point where the desired information to be learned can be gathered. If one desires to learn something about a characteristic fabricated on a grinding machine, data should be collected at the grinding operation. If one waits several operations after the fact before taking measurements, many opportunities to gain knowledge about the grinding operation may be lost. Waiting until final inspection to take measurements regarding in-process events is almost always too late to be of real-time benefit.

How Often to Collect Data

The frequency of data gathering depends on four factors.

1. The availability of data
2. The cost of gathering the data
3. The time interval between major process changes or adjustments
4. Process stability or uncertainty of process output

Take, for example, the autoclave data. The number of composite parts manufactured may not have been large, but the number of temperature measurements was large. That is, there were ample opportunities available to take temperature data. Also, because the data gathering was automated, the cost of data acquisition was likely to be low.

Because the data were inexpensive to take, temperatures were gathered every 15 seconds. The 25 subgroups represent only a few minutes worth of cure time. Within these few minutes there were no major changes or adjustments made to the process, its materials, computer programming, staffing, and so on. Additionally, the $\overline{X}$ and R chart proved to be in control. With this information in mind, one could comfortably conclude that the process did not significantly change over the 6.5 minute interval.

Because the process did not change, the frequency of data gathering could be decreased. That is, for the purpose of determining process control, taking a temperature reading every 15 seconds may be overkill. Generally, if a process does not change over some period of time, its frequency of data gathering can be reduced. The amount of decrease in frequency will depend on the comfort level of those working in the process as well as the risk, penalty, and/or costs involved in producing nonconforming product due to an undetected process change.

The key is to gather data frequently enough so that any important changes to the process are caught, but not so frequently that the data gathering itself is cost prohibitive. The question to be answered is, "How long can we go without sampling the process and still have confidence that the process is still consistent and that good product is still being produced?"

Sometimes, especially with new products or processes, time intervals between process changes are unknown. With new processes, one should generally take as much data as possible, even if one has to resort to gathering data on 100 percent of the products produced. In this way, process changes and their time intervals can be identified and the appropriate sampling frequency can be quickly established.

Frequency of data collection is typically a trade-off between minimizing data collection cost and maximizing the probability of detecting process changes. Once sampling frequency has been established, one must then consider the issue of subgrouping.

Determining Rational Subgroups

Generally, subgrouping strategies should be determined while keeping in mind these three important points.

1. Each subgroup should be as homogeneous as possible.
2. Each measurement in a subgroup must be independent of one another.
3. Using larger subgroups results in control charts that are more sensitive to shifts in the process mean.

The first point, *each subgroup should be as homogeneous as possible,* means that each subgroup should be composed of measurements that represent a similar time and place of manufacture. This usually means taking measurements of consecutively produced products at or from the same machine or location. Taking data in this manner ensures that the variation within the subgroup (demonstrated by plotted points on a range chart) and between subgroup ranges can be fairly compared with one another. Given this subgrouping strategy, short-term variability is illustrated by the range chart and longer-term process variability is illustrated by the $\overline{X}$ chart.

For example, a natural subgroup might be four consecutively produced inside diameters from machine 6, with subgroups taken four times during each shift. In this subgrouping plan, the sources of variation within a subgroup are known. Also, one could more easily problem solve and identify assignable causes of variability from shift to shift.

It would not be rational for the subgroup to be composed of two inside diameters from machine 1 on first shift, one from machine 3 on second shift, and one from machine 5 on third shift. If there were significant differences between shifts or machines, they would be very hard to identify. Not only would it be a headache to execute this subgrouping plan, it would prove difficult if not impossible to identify assignable causes of variation.

The second point, *each measurement in a subgroup must be independent of one another,* means that one measurement in a subgroup should not be influenced by another. Independence would be violated if a subgroup consisted of samples from each of five T/Cs placed evenly throughout an autoclave chamber. Different locations within the autoclave chamber are all dependent on the overall chamber temperature. A hot spot in one area of the chamber would affect other areas of the chamber. Therefore, the temperature measurements would be dependent on one another. To accurately represent process variability from all five chamber locations, one would have to use a separate traditional control chart to evaluate each T/C's temperature readings.

The third point to keep in mind when subgrouping is *using larger subgroups results in control charts that are more sensitive to shifts in the process mean.* Generally, subgroup sizes (n) range from 1 to 10, and sometimes larger. In the autoclave example, the subgroup size was 5. It could just as easily have been 3, 6, 10, or 14. Either way, the bigger the subgroup, the more sensitive the $\overline{X}$ chart is to changes in the average (given an unchanging process standard deviation). However, there is a trade-off. As the subgroup size, n, increases, the range (since it only uses two data points in a subgroup, the largest and smallest values) becomes less efficient as an estimator of process variability. Typically, when n is 10 or larger, the sample standard deviation s is a better statistic to use in estimating process variability since all of the data points in the subgroup are used in the s calculation.

Take, for example, a certain process with a known standard deviation of 2 and an average of 10. If one were using a subgrouping

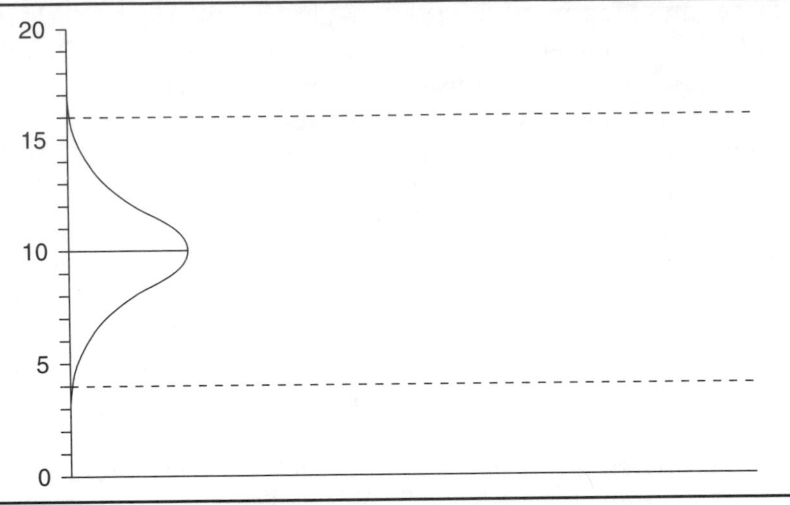

Figure 5.1. Distribution of individuals chart with $n = 1$.

scheme with $n = 1$, then the distribution of the individuals control chart would look something like Figure 5.1.

Figure 5.1 shows that the individual data points should normally vary between 4 and 16 (the values of the LCL and UCL, respectively). However, say that at time x the overall process average changes to 12, and at time y it changes again to 14. Assume that the estimate of the process standard deviation for the individual measurements remains the same as before ($\hat{\sigma} = 2$). Given these new circumstances, the location of the distribution of individual data points changes. A visual description of those changes can be found in Figure 5.2.

Based on the theory of the normal curve shown in Figure 5.3, one will find that, at time x, there is approximately a 2.5 percent chance of detecting the first change in the average. This is determined because the average has shifted so that the $+2\sigma$ boundary of the normal curve now falls on the UCL for the $\overline{X}$ chart.

In Figure 5.3, it is estimated that, for a normal distribution, approximately 95 percent of all process data falls between $\pm 2\sigma$; this leaves about 5 percent that falls outside of those bounds. Naturally, because only one tail of the curve is being evaluated, only half (2.5 percent) falls outside of the $+2\sigma$ boundary. In a similar fashion, the chance of detecting the second change in the process average (at time y) is approximately 16 percent.

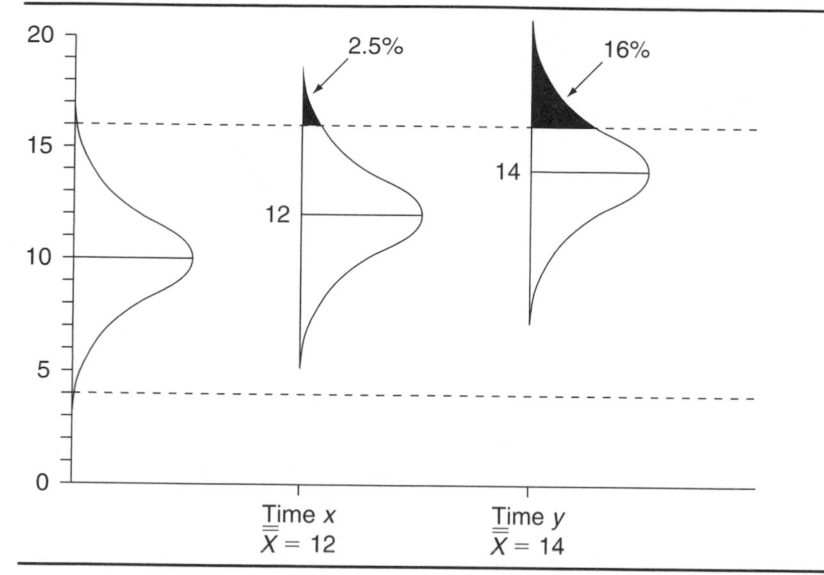

Figure 5.2. Chart showing distributions of individuals data with changes in overall process average.

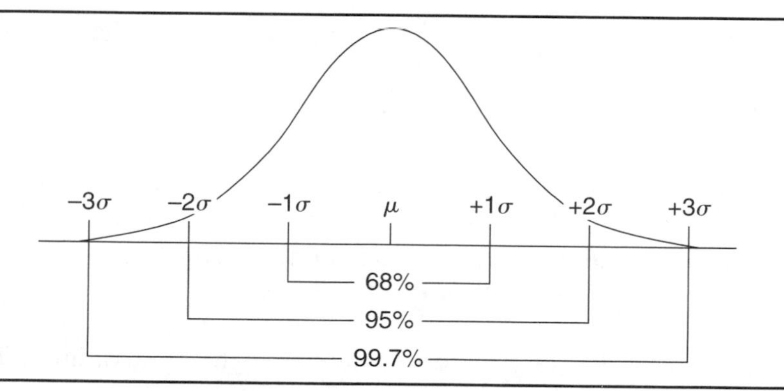

Figure 5.3. Areas of the normal curve.

Now consider the exact same process with the original average of 10 and standard deviation of 2. It is exactly the same situation except that an $\bar{X}$ chart is used with $n = 4$ instead of a chart whose control limits are based upon individuals data. Assume that the same changes in the overall process average occur at time x and at time y. The result is shown in Figure 5.4.

Notice that the control limits on the $\bar{X}$ chart are narrower than those on the individuals chart. This is because variation for averages

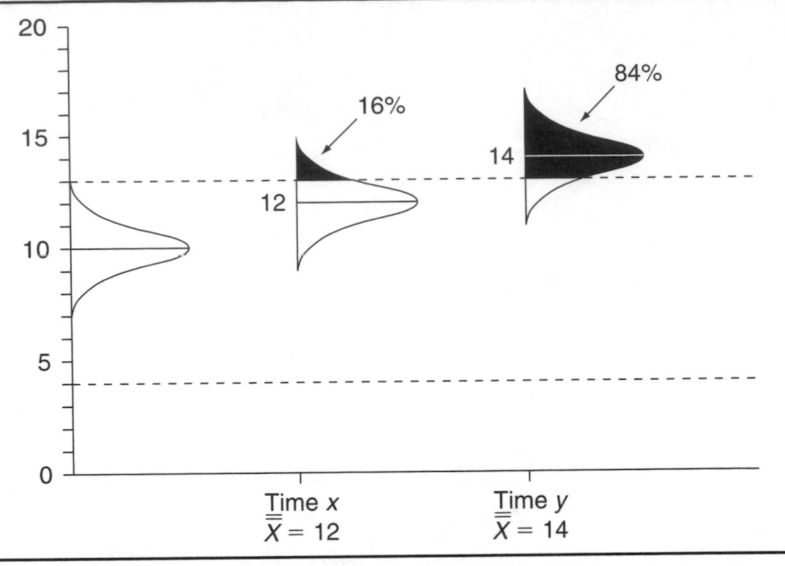

Figure 5.4. Chart of distributions of averages (n = 4) showing changes in the overall process average.

is always less than the variation for individuals. In fact, the standard deviation for averages is decreased by the inverse of the square root of n (see Calculation 5.1).

$$\hat{\sigma}_{\overline{X}} = \frac{\hat{\sigma}}{\sqrt{n}}$$

Calculation 5.1. Formula for estimating the standard deviation of averages.

The standard deviation used in calculating the control limits for the $\overline{X}$ chart is shown in Calculation 5.2.

$$\hat{\sigma}_{\overline{X}} = \frac{\hat{\sigma}}{\sqrt{n}} = \frac{2}{\sqrt{4}} = \frac{2}{2} = 1$$

Calculation 5.2. Estimating the standard deviation of averages.

Compare Figure 5.4 with Figure 5.2. Notice that, in Figure 5.4 when the process average changes to 12 at time *x*, there is approximately a 16 percent chance of detecting the shift with a subgroup size of four versus only a 2.5 percent chance with a subgroup of one (see Figure 5.2). Also, when the process average shifts to 14 at time *y*, there is approximately an 84 percent chance of detecting the shift with a subgroup size of four. Notice that, with a subgroup size of one, as in Figure 5.2, there is only a 16 percent chance of detecting the same change in the process average.

Keep in mind that the situations are similar; the only difference in the two scenarios is the size of the subgroups. These dramatic increases in the ability to detect changes in the average are due solely to using a larger subgroup size.

In summary, data collection strategies play a big role in the interpretation and use of control charts. Following the general rules outlined in this chapter should help to identify the most appropriate data collection strategy for most situations.

Chapter 6

Selecting the Right Chart

This chapter shows how to use the control chart decision tree to select the most appropriate chart for almost any situation (see Figure 6.1). To use the decision tree, only four questions need to be answered.

1. How many characteristics, parameters, or measurement locations from the same part or process are to be monitored on the same chart? The answer is either one or greater than one (1 or >1). For example, if a flange thickness is being measured at a single location on each part, the answer is one. However, if one wants to monitor flange thickness at several locations on the same part and track them all on the same chart, the answer would be greater than one (>1). For more details about multiple characteristics on the same chart, see chapter 19.

2. What is the subgroup size? There are three possible answers.

- One
- Greater than one but less than 10 (>1 but <10)
- Greater than or equal to 10 (≥ 10)

See chapter 5 for further discussion on how to select subgroup sizes.

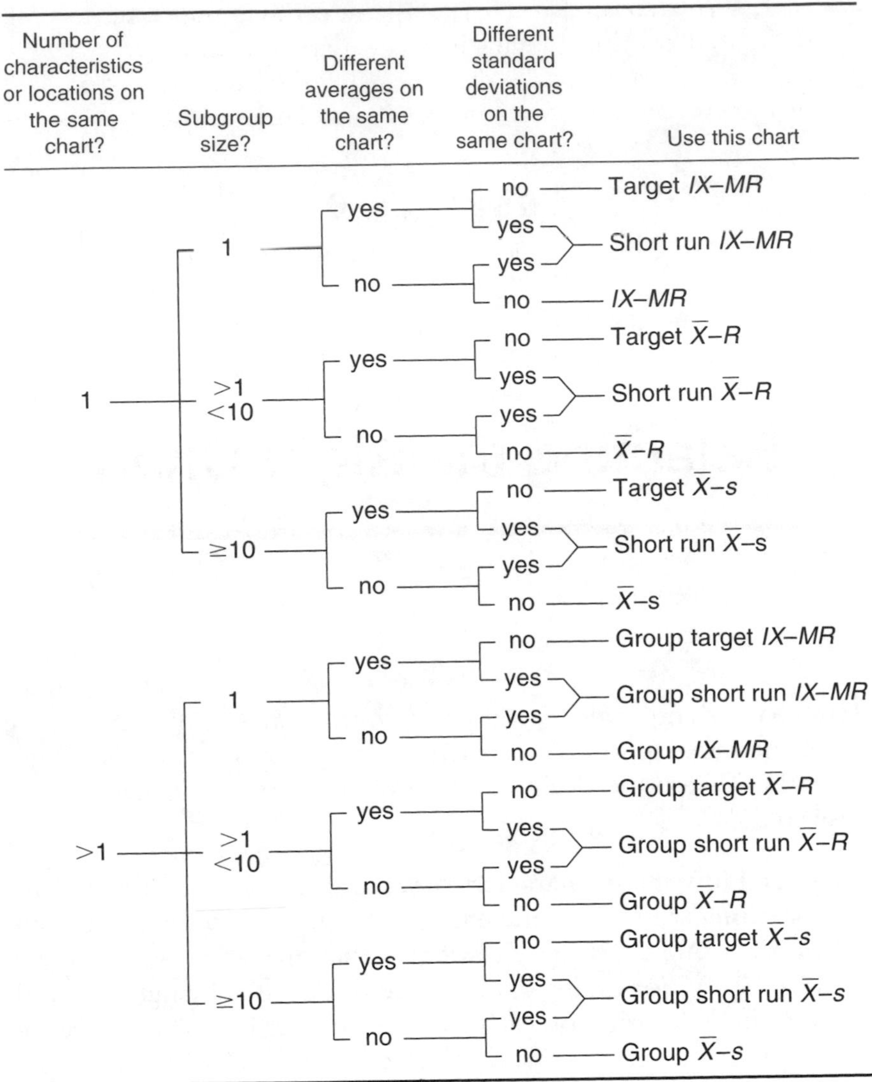

Key: IX = individual X chart; MR = moving range chart; R = range chart (R chart); s = s chart; $\bar{X}$ = $\bar{X}$ chart.

Figure 6.1. The control chart decision tree is used to determine which control chart is appropriate for a given situation.

3. Are characteristics with different averages going to be evaluated on the same chart? The answer is either *yes* or *no*. If a control chart is used to monitor many different sizes of outside diameters across all parts produced on a lathe, the answer would be *yes*. If the chart was going to be used to track only 0.500-inch diameters, the answer would be *no*.

4. Are characteristics or measurement locations with different standard deviations going to be evaluated on the same chart? The answer is either *yes* or *no*. If a control chart is used to monitor hole diameters produced by a single machine, with some holes finished with a drill while others with a reamer, the standard deviations would likely be different and the answer would be *yes*. If all holes, regardless of size, were finished with a reamer, the answer would probably be *no*.

If no knowledge exists about the standard deviations, assume them to be similar and let the data show otherwise. The standard deviation of a suspect characteristic can be assumed to be different if the difference is more than 30 percent of the average standard deviation (or range) of the other parts on the chart. (Note that this 30 percent difference is an estimate and should serve only as a rule of thumb.) Other statistical methods, such as the F-test or Kruskal-Wallis (Kw) test are also used to detect differences in standard deviations. These methods, however, will not be covered in this text.

Example 1

A lathe is used to produce outside diameters (ODs) of different sizes. Several types of alloy material are run through this lathe. To select a single control chart to monitor all ODs regardless of size or material, the following questions from the decision tree are answered.

1. *How many characteristics or locations are on the same chart?*

Only one outside diameter is measured per part because only one measurement is required to verify all diameters across the part.

2. *What is the subgroup size?*

Three is selected because the short-term variation in the lathe tends to change at different times of the day. The machinist wants adequate visibility of these changes using a minimum number of measurements per subgroup.

Number of characteristics or locations on the same chart?	Subgroup size?	Different averages on the same chart?	Different standard deviations on the same chart?	Use this chart
1	>1 <10	Yes	Yes	Short run $\bar{X}$–R

Figure 6.2. Progression through control chart decision tree for example 1.

3. *Are there different averages on the same chart?*

Yes. Each new part being run usually has a different diameter than the last.

4. *Are there different standard deviations on the same chart?*

Yes. Softer material usually exhibits less variation. With material of varying degrees of hardness being turned, the standard deviations are likely to be different.

Following the responses through the decision tree, the correct chart to use is a *short run* $\bar{X}$ and range™ chart. This progression of questions through the control chart decision tree can be found in Figure 6.2.

Example 2

A conventional mill is set up to cut only 0.250" radii. Different but similar parts are machined on this mill. There are three radii on each part being measured. Referring to the control chart decision tree, the answers to the following questions lead to the most appropriate chart for the situation.

1. *How many characteristics or locations are on the same chart?*

The milling operation cuts three separate radii on each part.

2. *What is the subgroup size?*

One is selected because each run is usually less than 10 parts. Also, history has shown that the greatest amount of variation occurs between jobs.

3. *Are there different averages on the same chart?*

No. The mill only cuts 0.250-inch radii.

Number of characteristics or locations on the same chart?	Subgroup size?	Different averages on the same chart?	Different standard deviations on the same chart?	Use this chart
>1	1	No	No	Group *IX–MR*

Figure 6.3. Progression through control chart decision tree for example 2.

4. *Are there different standard deviations on the same chart?*
No. All parts are made from the same material.

Following the responses on the control chart decision tree, the correct chart to use is a group individual *X* and moving range chart (see Figure 6.3).

Example 3

A heading machine is used to create a finished head height dimension on fasteners. An automatic device is used to measure all head heights immediately after being ejected from the die. For SPC purposes, 10 consecutive measurements every 15 minutes are automatically downloaded into SPC software. Many different part numbers are run on this machine and typical job runs are 30,000 parts. The decision tree is consulted to select the best control chart for the heading machine.

1. *How many characteristics or locations are on the same chart?*
One. Only head heights are being measured.
2. *What is the subgroup size?*
Ten is selected because data are readily available and inexpensive to collect. Also, the short-term variation of the head heights helps the machinist evaluate the stability of the setup.
3. *Are there different averages on the same chart?*
Yes. Head heights are different for every job.
4. *Are there different standard deviations on the same chart?*
No. All fasteners are made from similar material.

Number of characteristics or locations on the same chart?	Subgroup size?	Different averages on the same chart?	Different standard deviations on the same chart?	Use this chart
>1	≥10	Yes	No	Target $\overline{X}$–s

Figure 6.4. Progression through control chart decision tree for example 3.

Following the responses on the control chart decision tree, the correct chart to use is a target $\overline{X}$ and s chart (see Figure 6.4).

Every control chart application is different. Each data collection strategy requires careful consideration. Always remember: first define the questions about the process, then choose the right control chart to reveal the answers.

Chapter 7

Traditional Control Charts for Single Characteristics

In this section, three well-known, traditional control charts will be covered.

- Individual X and moving range chart
- $\overline{X}$ and range chart
- $\overline{X}$ and s chart

These charts are the ones with which most SPC practitioners have their experience. They are also the very backbone of this text and are the foundation upon which all other control charts in this book have been built. This section covers these basic, traditional control charts in detail. In addition, information on how to use and maintain these traditional charts is covered, as well as when and when not to recalculate control limits.

These traditional charts are most appropriately used in situations where the production runs are long and rarely changing, data are plentiful and inexpensive to gather, and there is only one or a very small number of characteristics to be evaluated. Traditional control charts were developed to be applied under these specific circumstances.

For each of the three traditional control charts, notice that only one item defines their difference: subgroup size n. If the control chart user wishes to place a dot on the chart each time a measurement is taken, the individual X and moving range chart is appropriate. If the user defined the subgroup size to be greater than one but less than 10, the $\overline{X}$ and range chart would be used. If one had enormous amounts of inexpensive-to-gather data and defined the subgroup size to be 10 or larger, the $\overline{X}$ and s chart would be the chart to use.

Once the type of control chart is selected, the user generates a control chart for each quality characteristic. For example, say two different characteristics were of interest—the outside and inside diameters of aluminum piping. The subgroup size is five. The user would gather five inside diameters and calculate both an $\overline{X}$ and an R value. The subgroup values for $\overline{X}$ and R would then be placed in the form of plot points on each respective chart. The operator would repeat the procedure for the five OD values and place the plot points on a chart *separate* from the inside diameters. One chart for the inside diameters, one for ODs. If three characteristics were of interest, the SPC user would need three separate $\overline{X}$ and R charts, and so on.

Using Traditional Control Charts

Once the subgroup size is defined and a control chart has been selected, it is just a matter of putting plot points on a chart, right? Well no, it's not quite that easy. Selecting the appropriate chart is only one of the number of challenges facing the SPC practitioner. One difficult issue is the management of the charts once they are in use.

In managing control charts, SPC practitioners will be interested in identifying assignable causes and either eliminating them (if they have a detrimental effect on process performance) or incorporating them into the process (if the cause enhances performance). This is what is meant by *process control*—searching out reasons for changes in process performance and reacting to those changes. Control charts are a means for controlling the predictability and performance of characteristics from a process.

Process control using a control chart has two different stages: establishing baseline data and control limits and maintaining the control chart.

Establishing Baseline Data and Control Limits

First, one needs to establish a baseline from which to work. To do this, one should identify an appropriate control chart to use, define the subgroup size, and begin to gather data. Once 15 to 25 subgroups have been gathered, control limits can be calculated. If the

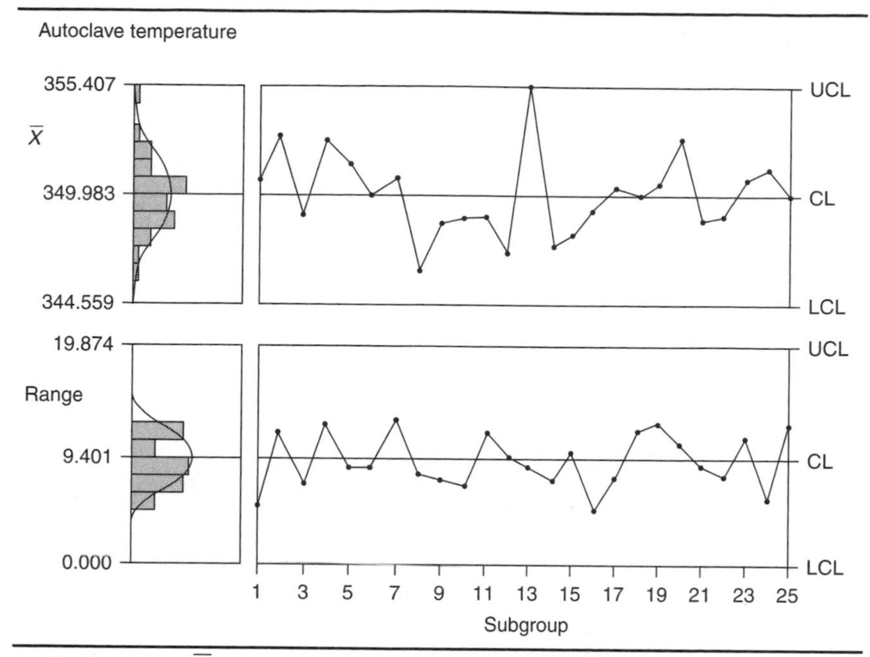

Figure 7.1. $\overline{X}$ and range chart of autoclave temperature readings.

chart shows a lack of control, then one should problem solve, identify, and then eliminate assignable causes in the process. Once assignable causes have been eliminated, the revised control chart should be in control. This chart should then serve as the process' baseline.

Take, for example, the data from the autoclave in Table 2.1. In our example, temperature readings were gathered every 15 seconds from a single thermocouple (T/C) in the autoclave. The $\overline{X}$ and range chart found in Figure 7.1 shows the temperature readings to be in control.

Recall that we determined the temperatures were in control with respect to both the standard deviation and average of those temperatures. This $\overline{X}$ and R chart will serve not only as a baseline for identifying assignable causes in the process, but will also serve as the control chart by which process improvements and changes will be evaluated.

Maintaining the Control Chart

The second step in using a control chart is to maintain it. Once the baseline has been established, the chart user should take the baseline control limits and extend them into the future (see Figure 7.2).

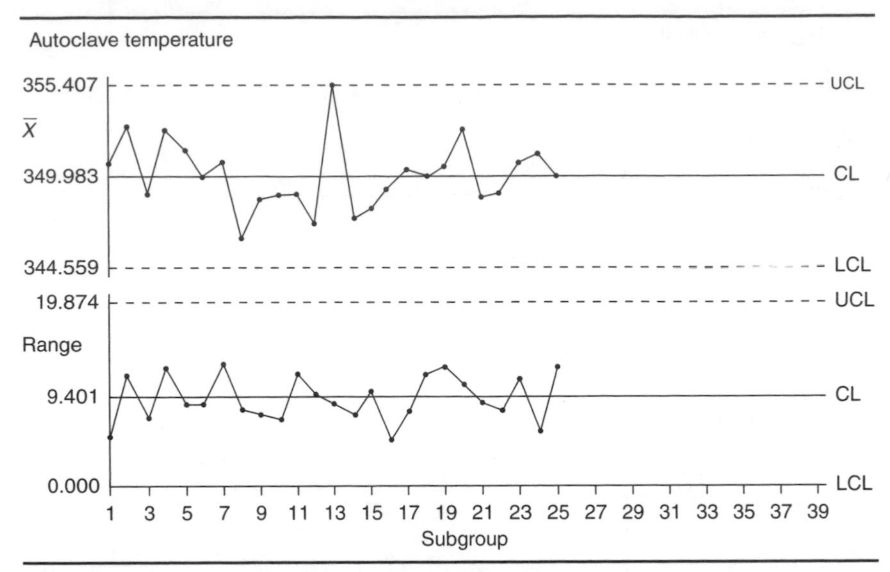

Figure 7.2. Baseline control limits extended into the future.

Extending control limits into the future allows one to judge subsequent temperature data against the established baseline. If the temperatures change significantly, the change will show up as an out-of-control signal (see Figure 7.3). In Figure 7.3, assume that more temperature readings have been taken from the autoclave and new data have been plotted on the baseline control chart.

These new data added to the chart suggest that the standard deviation for temperature values has decreased significantly. This conclusion is reached because there is a run of nine range values below $\overline{R}$. The run of nine represents the presence of an assignable cause.

Notice that the $\overline{X}$ chart seems to be in control. However, we cannot make this conclusion because the old $\overline{X}$ chart limits are no longer appropriate. Recall from chapter 3 that the control limits for the $\overline{X}$ chart are calculated using the $\overline{R}$ value from an in-control range chart. Because the range chart reveals an unstable condition (nine consecutive points in a row below the centerline), this means that the control limits for the $\overline{X}$ chart will have also changed. Therefore, it is inappropriate to speculate on the control of the $\overline{X}$ chart at this point.

What is known, though, is that the standard deviation in temperatures has decreased significantly. The next step would be for the chart user to identify the assignable cause associated with this decrease in temperature variability.

Some might mistakenly argue that, because the reduction in temperature variation represents an improvement, time would be wasted

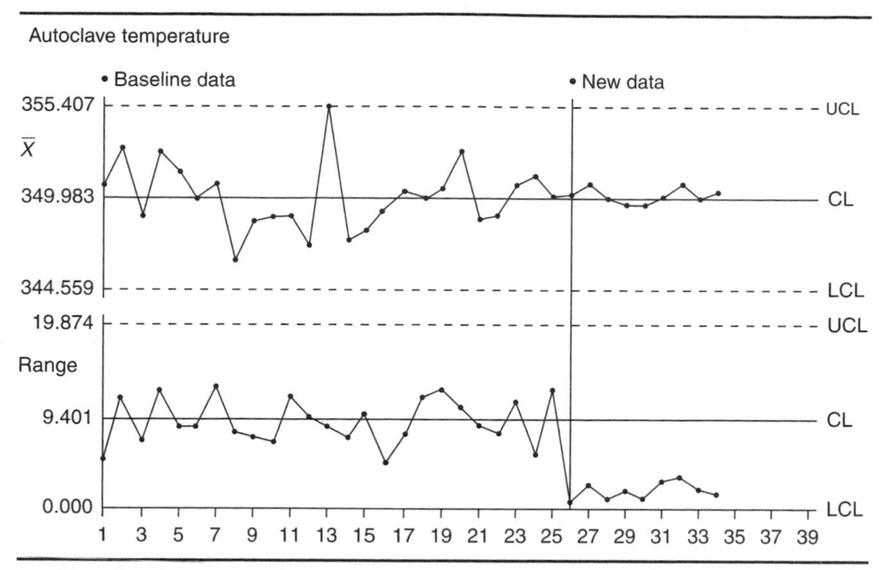

Figure 7.3. Autoclave control charts with extended control limits and nine new plot points.

trying to identify the assignable cause. This would truly be a mistake. SPC practitioners must not only identify process problems, but also opportunities. These new data may represent an opportunity for improvement that could be used in other autoclaves or a new learning that could be incorporated permanently into the present autoclave.

After looking around and asking questions, the chart user discovered engineers had installed a new temperature controller for the autoclave. Therefore, these data reflect an improvement in the control of the temperature as compared with the old controller. This is very good news. (Recall that the process was not capable because its C_p value was less than 1.) It means that the process is more capable than before the installation of the new temperature controller. It reflects a reduction in variability of the temperatures for the T/C being evaluated.

One could conclude that the assignable cause associated with the reduction in temperature variation was the new controller. If this is, in fact, the assignable cause, one might expect the reduction in temperature standard deviation to be long lasting, if not permanent. Therefore, one would also expect that, in the future, more temperature values should fall within the engineering specification limits, translating into less scrap and rework. This certainly represents a significant improvement to the process.

However, the chart user is now faced with the problem that the old baseline control limits are just that—old. They do not represent the current operation of the process and the change of the temperature controller. The control limits that have been extended into the future are no longer reflective of the current process performance. The control chart user is faced with a decision: what should be done with old control chart data and new improvements to the process?

The answer is that new control limits should be calculated for the process. Recalculation of control limits can be confusing when maintaining control charts. When to recalculate and when not to recalculate control limits is not always very clear. In his book, *Understanding Statistical Process Control*, Donald J. Wheeler writes, "The revision of control limits should be considered only when there is reason to suspect that the current limits are not appropriate . . . the obvious time to consider a revision of the control limits is when the process has been changed."[1] In other words, recalculating control limits should be done only when the current limits do not accurately reflect the present operation of the process.

Generally, control limit recalculation should be considered when all four of the following items are found to be true.

1. The process has changed. (Indications of assignable causes, especially a shift, are present.)
2. The cause of the process change is known.
3. The process change is expected to continue.
4. There are enough data to recalculate (15 to 25 plot points).

For the autoclave example, the first three items are true. First, we know the process has changed because there was a run of nine range values below the $\overline{R}$. Second, the cause of the process change was identified as a newly installed temperature controller. Third, one would expect the temperature improvements to continue because inclusion of a new temperature controller is a permanent hardware change. However, the fourth item is still in question.

One traditionally needs about 15 to 25 plot points to calculate control limits. Since there already exists nine new plot points, one might be tempted to just take the six points prior to the controller change and use those with the new data points to create new control limits. This would be a mistake. Mixing old data with current data would result in control limits still not reflective of the current operation of the process. Instead, the correct thing to do would be to wait until an additional six or more subgroups of the new data are acquired and then recalculate control limits.

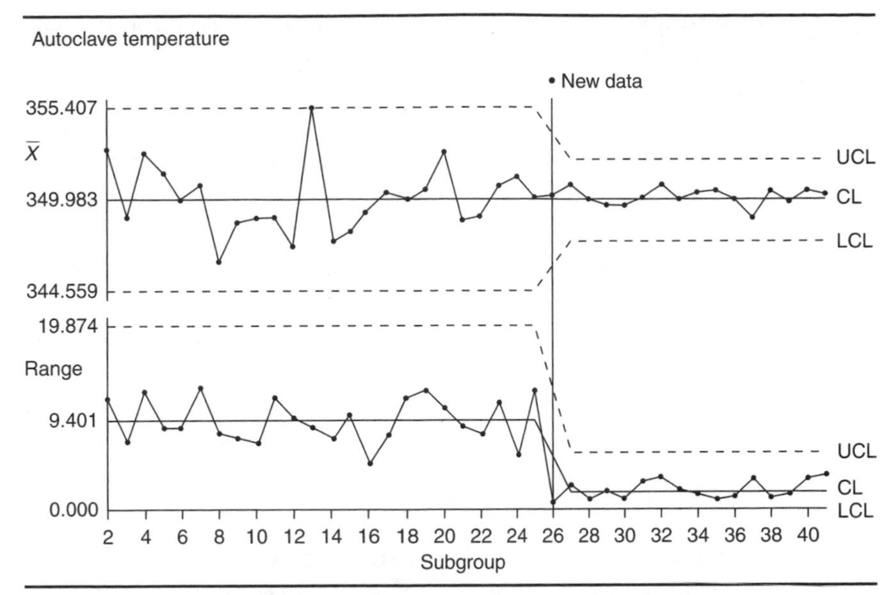

Figure 7.4. Autoclave control charts with new baseline control limits and data.

Doing so results in a control chart that looks like Figure 7.4.

The new control limits become the new baseline for the autoclave temperatures. Maintenance of the control chart should continue as discussed earlier. Notice that, not only has a new baseline been established, but the old and new data can be used as proof of process improvement activities. The chart can be used as reinforcement to operators, managers, and engineers alike that hardware changes that have been implemented resulted in significant improvements to the process.

As important as it is to know when to recalculate control limits, it is equally important to know when *not* to recalculate control limits. For traditional control charts, limits should not be recalculated

1. When there are less than 15 subgroups of data. Say someone gathered two subgroups of data and then calculated control limits from those data. Because control limits are predictions of future process behavior, most people would conclude that two subgroups are just not enough data to make a good, believable prediction. In effect, using less than 15 subgroups to calculate control limits is tantamount to calculating control limits with only

two subgroups—there is simply not enough information to make a reliable prediction of future process behavior.

2. With more than 40 subgroups of data. The first instance represents a situation in which not enough data exist, but this is a situation of too much data. Using more than 40 subgroups to calculate control limits may represent too long of a period of time in a process. Process changes that might occur within the 40 subgroup time frame may unnaturally inflate the estimate of the control limits.

3. With each new subgroup of data. Recalculating control limits after every additional subgroup is gathered may mask assignable causes in a process. Say that, in the autoclave example, there is a very gradual decrease of T/C temperatures over time. This gradual decrease over time should reveal a trend downward as the $\overline{X}$ chart shows in Figure 7.5.

Figure 7.5 represents a correctly constructed control chart. If, however, control limits are recalculated after each subgroup, the gradual change in the process would be incorporated into each new set of control limits. This could result in a situation where subgroup averages would not fall below the LCL. Therefore, the control chart would not show indications of assignable causes (except for the trend pattern) and the process operators might not be alerted to the decreasing hole diameters unless they were to look at the historical trend on the entire chart.

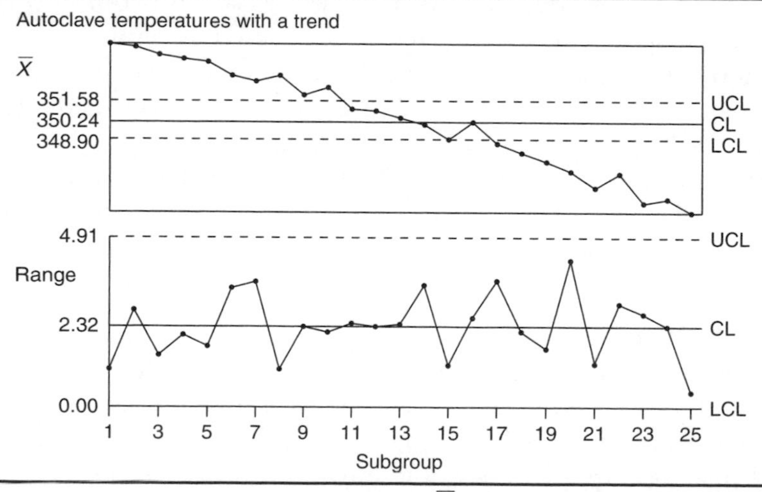

Figure 7.5. Autoclave temperatures $\overline{X}$ chart with downward trend and an in-control range chart.

This situation is not usually encountered by SPC practitioners who rely on paper control charts. Instead, it is usually found when those practitioners rely on software. We have seen several SPC software packages in which the program is written to recalculate control limits with every additional subgroup of data. Therefore, one would be wise to check the software manual to determine not only how control limits are calculated, but the frequency of their recalculation.

Note

1. Donald J. Wheeler, *Understanding Statistical Process Control,* 1st ed. (Knoxville, Tenn.: SPC Press, Inc., 1986), 78.

Chapter 8

Individual *X* and Moving Range Chart

Decision Tree Section

Table 8.1. Control chart decision tree.

Number of characteristics or locations on the same chart?	Subgroup size?	Different averages on the same chart?	Different standard deviations on the same chart?	Use this chart
1	1	No	No	*IX–MR*

Description

Individual X *Chart*

The individual *X* chart, also called the *IX* or just *X chart,* is used to monitor and detect changes in the process mean by evaluating the consistency of individual measurements of a single characteristic. Because the plot points represent individual measurements, the subgroup size is one. *IX–MR* charts are intended to be used to monitor characteristics where only one measurement can represent the process at a given period of time. Examples include monthly financial values or homogeneous batches of a manufactured product such as concentration in a chemical bath.

Moving Range Chart

The *moving range* chart (also called an *MR chart*) is used to monitor and detect changes in the standard deviation of individual measurements from a process. The plot points represent the absolute difference between two consecutive individual measurements. Although not the same as standard deviation, the *MR* values can be used to estimate the process standard deviation.

Subgroup Assumptions

- Independent measurements
- Constant sample size, n = 1
- One characteristic
- One unit of measure
- Measurements normally distributed

Calculating Plot Points

Table 8.2. Formulas for calculating plot points.

Chart	Plot point	Plot point formula
Individual *X*	*IX*	Each individual measurement
Moving range	*MR*	Absolute difference between two consecutive *IX* values

Calculating Centerlines

Table 8.3. Formulas for calculating centerlines.

Chart	Centerline	Centerline formula
Individual *X*	$\overline{IX}$	$\frac{\sum IX}{k}$
Moving range	$\overline{MR}$	$\frac{\sum MR}{k-1}$

Calculating Control Limits

Table 8.4. Formulas for calculating control limits.

Chart	Upper control limit	Lower control limit
Individual *X*	$\overline{IX} + A_2\overline{MR}$	$\overline{IX} - A_2\overline{MR}$
Moving range	$D_4\overline{MR}$	0

Example

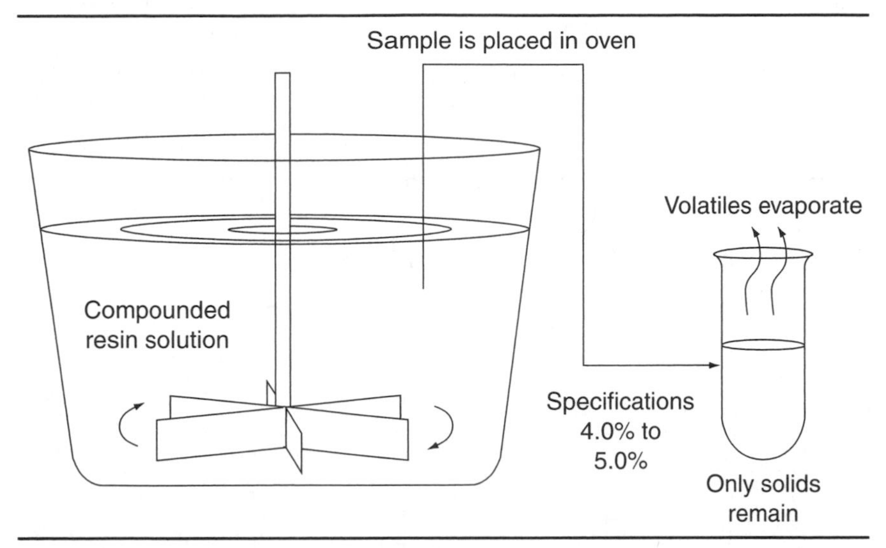

Figure 8.1. Solids from a solution are measured by taking a sample of known weight and evaporating the volatiles. The percent difference in weight of the remaining powder is the "percent solids."

Case Description

Solids content of a compounded resin solution is a key characteristic. To obtain the percent solids, a sample of known volume is taken from the resin solution. The sample is then baked until all volatiles are evaporated and only the solids remain. The solids are weighed and a percent of total volume is calculated.

Sampling Strategy

Because only one characteristic is being controlled and only one measurement represents each batch, the *IX–MR* chart was selected. One measurement is taken from each batch.

Data Collection Sheet

Table 8.5. Data collection sheet for percent solids *IX* and *MR* chart.

	1st shift					2nd shift				
k	1	2	3	4	5	6	7	8	9	10
IX	5.2	4.9	5.0	5.1	4.8	4.1	3.9	4.0	4.2	4.1
MR	—	0.3	0.1	0.1	0.3	0.7	0.2	0.1	0.2	0.1

Table 8.5. Cont.

	1st shift					2nd shift				
k	11	12	13	14	15	16	17	18	19	20
IX	4.9	5.0	4.8	5.0	5.1	4.0	4.3	4.3	4.4	4.1
MR	0.8	0.1	0.2	0.2	0.1	1.1	0.3	0.0	0.1	0.3

Individual X and Moving Range Chart

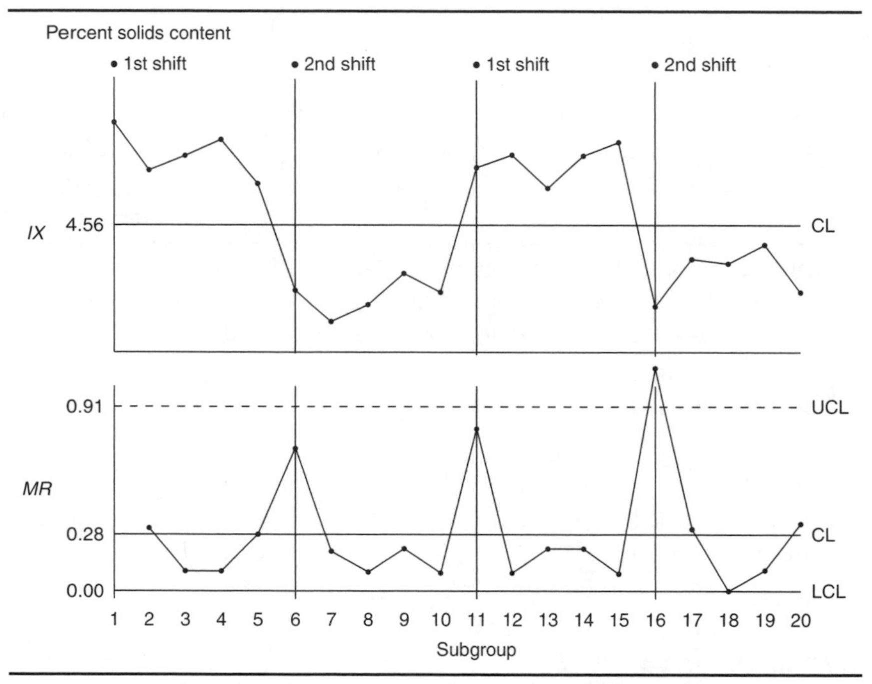

Figure 8.2. Percent solids *IX* and *MR* chart.

Calculations for the MR *Chart*

$$\overline{MR} = \frac{\sum MR}{k-1} = \frac{5.30}{19} = 0.28$$

$$\text{UCL}_{MR} = D_4\overline{MR} = 3.267(0.28) = 0.91$$

$$\text{LCL}_{MR} = 0$$

Calculation 8.1. Calculations for *MR* chart.

MR *Chart Interpretation and Recalculation*

Upward spikes occur at every shift change on the *MR* chart. This is an indication that the variation in individual data values increases significantly at those points. Looking closely at the data, one will find that these increases in variation are likely due to mean percent solids differences between shift 1 and shift 2.

Because the *MR* chart is out of control, the value of $\overline{MR}$ is unreliable and cannot be used to calculate control limits for the *IX* chart. This is why no control limits were placed on the *IX* chart in Figure 8.2.

To obtain a reliable estimate of $\overline{MR}$ and control limit values for the *IX* chart, one must

1. Remove the out-of-control plot points from the *MR* chart.
2. Recalculate $\overline{MR}$ with the remaining MR values.
3. Determine if the remaining *MR* values are in control with respect to the new control chart calculations.

The recalculations for the *MR* chart are performed in Calculation 8.2.

$$\overline{MR} = \frac{\Sigma MR}{k-1} = \frac{2.70}{16} = 0.17$$

$$\text{UCL}_{MR} = D_4\overline{MR} = 3.267(0.17) = 0.56$$

$$\text{LCL}_{MR} = 0$$

Calculation 8.2. Revised *MR* chart calculations (after removing out-of-control data values for subgroups 6, 11, and 16).

Note that all of the 16 remaining moving range values fall within the new *MR* chart control limits (see Figure 8.3). There appears to be no indication of assignable causes of variation. Given this situation, it is now appropriate to complete the control chart calculations for the *IX* chart.

Calculations for the IX *Chart*

$$\overline{IX} = \frac{\Sigma IX}{k} = \frac{91.2}{20} = 4.56$$

$$\text{UCL}_{IX} = \overline{IX} + A_2\overline{MR} = 4.56 + 2.660(0.17) = 5.01$$

$$\text{LCL}_{IX} = \overline{IX} - A_2\overline{MR} = 4.56 - 2.660(0.17) = 4.11$$

Calculation 8.3. Calculations for *IX* chart.

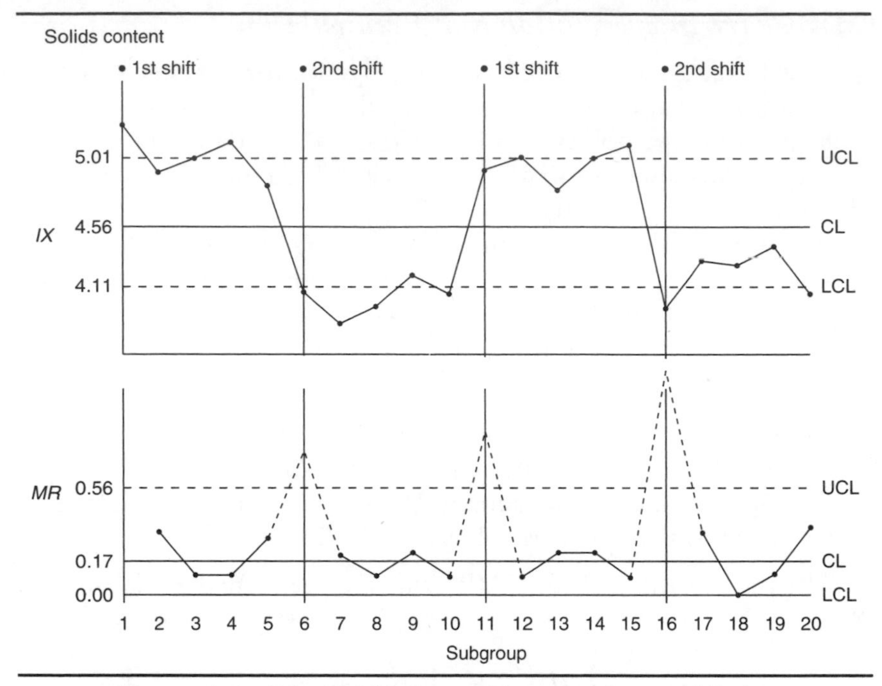

Figure 8.3. *IX* and *MR* chart with revised control limits. Subgroup numbers 6, 11, and 16 have been removed from calculations for the *MR* chart.

IX *Chart Interpretation*

The *IX* chart shows a lack of control because several points fall outside of control limits (see Figure 8.3). Also, there is a definite pattern in the plot points. All first shift plot points fall above the centerline and all second shift plot points fall below. It seems as though first shift produces resin batches with a higher average solids content than second shift.

Recommendation

The variability within each shift appears to be similar. If both shifts could make adjustments so that they shared the same target for percent solids content, the shift-to-shift output would be more uniform and the potential long-term variability would be drastically less.

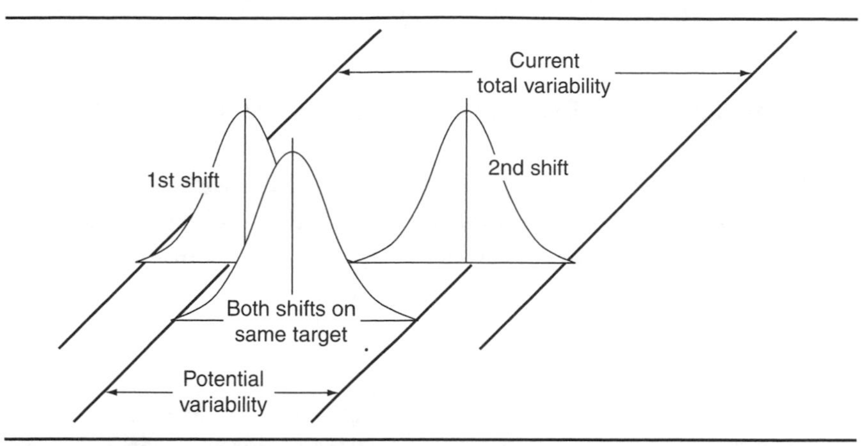

Figure 8.4. The total variability is largely due to the difference in means between the two shifts. Total long-term variability should be reduced when the two shifts share the same target.

Estimating the Process Average

The *IX* chart is not in control. Keep in mind that the $\overline{IX}$ on the control chart neither represents shift 1 nor shift 2. Instead, it represents the overall average solids content from both shifts. Therefore, it would be inappropriate to estimate the process average using all of the data from the current *IX* chart.

However, one could estimate the process average for each shift separately. This would be appropriate if

1. Enough data could be gathered from each shift to construct control charts.
2. The control charts were in control and showed no assignable causes of variation.

Of course, obtaining a reliable estimate for each shift's overall average would be done only for the purpose of calculating C_{pk} values for each shift. The most important action that can be taken is to remove the assignable causes responsible for the significant difference in percent solids between the two shifts.

If it can be assumed that the assignable causes have been removed, then one can estimate the process average for each shift. For illustration purposes only, the process average has been estimated in Calculation 8.4 for shift 1.

$$\overline{X} = \frac{\Sigma X}{k} = \frac{49.8}{10} = 4.98$$

Calculation 8.4. Estimate of the process average for shift 1.

Recall that the estimate of the process average for shift 1 is calculated using only 10 data points. Because of this relatively small sample of data, the estimate of the process average may not be reliable.

Estimating σ

To estimate σ, one first needs a reliable value for $\overline{MR}$. Recall that the $\overline{MR}$ and control limits were recalculated after removing assignable causes in Figure 8.2. The new *MR* chart was found to be in control for the remaining data values (see Figure 8.3). The new $\overline{MR}$ was calculated as shown in Calculation 8.5 and can be used in estimating the process standard deviation (Calculation 8.6).

$$\overline{MR} = \frac{\Sigma MR}{k-1} = \frac{2.70}{16} = 0.17$$

Calculation 8.5. Calculation of $\overline{MR}$ (to be used in estimating process standard deviation).

$$\hat{\sigma} = \frac{\overline{MR}}{d_2} = \frac{0.17}{1.128} = 0.15$$

Calculation 8.6. Estimate of the process standard deviation using the recalculated $\overline{MR}$.

Calculating Process Capability and Performance Ratios

Since there now exists a reliable estimate of the process standard deviation, it is appropriate to calculate the process capability ratio (C_p) as shown in Calculation 8.7.

$$C_p = \frac{\text{USL} - \text{LSL}}{6\hat{\sigma}} = \frac{5.0 - 4.0}{6(0.15)} = \frac{1.0}{0.9} = 1.11$$

Calculation 8.7. Calculation of the process capability ratio, C_p.

Recall that the IX chart demonstrated a lack of control. Therefore, the calculated $\overline{IX}$ for all 20 data points cannot be used as a reliable estimate of the overall process average. For this reason, it would be inappropriate to calculate an overall C_{pk} value using all 20 data values.

Instead, what one might do is separate the shift 1 IX data from the shift 2 IX data, construct separate IX and MR control charts, and confirm their consistency. Then, C_{pk} values could be calculated separately for each shift.

In order to illustrate the calculation of C_{pk} values for an IX and MR chart, the $\overline{IX}$ value from Calculation 8.4 will be used to calculate C_{pk} values representing shift 1 only.

$$C_{pk_u} = \frac{\text{USL} - \overline{IX}}{3\hat{\sigma}} = \frac{5.00 - 4.98}{3(0.15)} = \frac{0.02}{0.45} = 0.04$$

Calculation 8.8. $C_{pk\ upper}$ calculation using percent solids data from shift 1.

$$C_{pk_l} = \frac{\overline{IX} - \text{LSL}}{3\hat{\sigma}} = \frac{4.98 - 4.00}{3(0.15)} = \frac{0.98}{0.45} = 2.18$$

Calculation 8.9. $C_{pk\ lower}$ calculation using percent solids data from shift 1.

Because the number of IX measurements used for calculating the process average for shift 1 is only 10, the C_{pk} calculations are shown for illustration purposes only.

Notice that the C_p value is larger than 1, indicating that the process is capable of producing percent solids within the specification of 4.0 percent to 5.0 percent. However, the C_{pk} value of 0.04 means that, while the process is capable, it is not performing to specifications. This conclusion can be made for shift 1 only. Process capability and performance values for shift 2 can be found at the end of the chapter.

IX–MR *Chart Advantages*

- Easy to understand.
- The only control chart (IX) on which specification limits can be placed.

- Only 15 to 25 individual measurements are necessary to estimate control limits.
- Data can be plotted after each individual measurement is obtained.
- Minimal calculations.

IX–MR *Chart Disadvantages*

- Does not independently separate variation in the average from variation in the standard deviation.
- The histogram of the individual measurements must be approximately normal for the control limits to accurately represent $\pm 3\sigma$ limits.
- Not sensitive enough to quickly identify small changes in the process average or standard deviation.

Additional Comments About the Case

- The only reason specification limits may be placed on an *IX* chart is because the plot points are the actual individual measurements. Placing specification limits on a control chart, however, could prove to be confusing to the uninitiated user of control charts. Caution should be exercised if one chooses to place specifications on a control chart.
- Shift differences illustrated in this example are not uncommon in industry. Resolving these differences can sometimes be difficult due to rivalries or politics between shifts.
- The estimate of the process average and process capability and performance ratios for percent solids for shift 2 are shown in Table 8.6.

Table 8.6. C_p and C_{pk} values for percent solids produced by shift 2.

Shift 2
$\overline{IX}_2 = 4.14$
$C_{p2} = 1.11$
$C_{pk_u 2} = 1.91$
$C_{pk_l 2} = 0.31$

Chapter 9

$\overline{X}$ and Range Chart

Decision Tree Section

Table 9.1. Control chart decision tree.

Number of characteristics or locations on the same chart?	Subgroup size?	Different averages on the same chart?	Different standard deviations on the same chart?	Use this chart
1	>1 <10	No	No	$\overline{X}$ and range

Description

$\overline{X}$ *Chart*

The $\overline{X}$ chart is used to monitor and detect changes in the average of a single measured characteristic. The plot points represent subgroup averages. The subgroup size may range from two to nine, however, practitioners generally use subgroup sizes of three or five.

Range Chart

The range chart, also called an *R chart,* is used to monitor and detect changes in the standard deviation of a single measured characteristic. The plot points represent subgroup ranges. Ranges are not the same as standard deviation, but when subgroup sizes are small—less than 10—ranges can be used to closely estimate within subgroup standard deviation.

Subgroup Assumptions

- Independent measurements
- Constant sample size
- One characteristic
- One unit of measure

Calculating Plot Points

Table 9.2. Plot point formulas for the $\overline{X}$ and range chart. Subgroup size is represented by *n*.

Chart	Plot point	Plot point formula
$\overline{X}$	$\overline{X}$	$\frac{\Sigma X}{n}$
Range	R	Largest X – Smallest X (in a subgroup)

Calculating Centerlines

Table 9.3. Control chart centerline formulas for the $\overline{X}$ and range chart. The number of subgroups is represented by *k*.

Chart	Centerline	Centerline formula
$\overline{X}$	$\overline{\overline{X}}$	$\frac{\Sigma \overline{X}}{k}$
Range	$\overline{R}$	$\frac{\Sigma R}{k}$

Calculating Control Limits

Table 9.4. Control limit formulas for the $\overline{X}$ and range chart.

Chart	Upper control limit	Lower control limit
$\overline{X}$	$\overline{\overline{X}} + A_2\overline{R}$	$\overline{\overline{X}} - A_2\overline{R}$
Range	$D_4\overline{R}$	$D_3\overline{R}$

Example

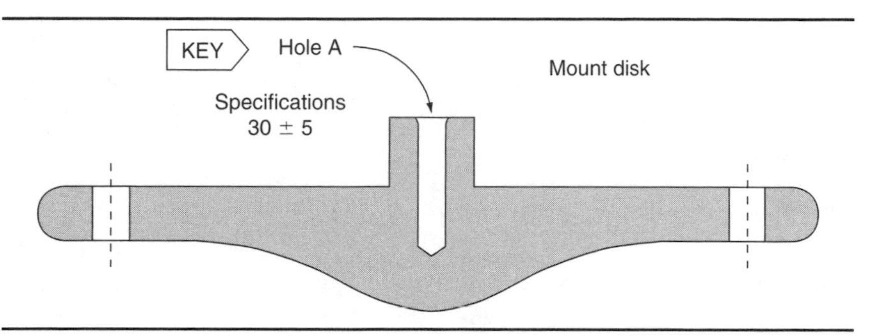

Figure 9.1. Mount disk with a single key characteristic, hole A.

Case Description

The inside diameter of hole A is a key characteristic. The customer has complained that the diameter is not consistent. There are no other key characteristics on the part.

Sampling Strategy

Because only one characteristic is being controlled and the production volume is high, an $\overline{X}$ and range chart was selected. To determine how often measurements should be taken, the machinist was consulted. With the goal of maximizing tool life, it was discovered that an oversized drill is used—one that cuts the hole on the high side of the tolerance—and the drill is replaced every couple of hours (about 500 parts). As the tool wears, the hole gets smaller. With this information it was decided that three measurements should be taken every 15 minutes. This would generate about eight plot points between drill replacements—enough to detect a tool wear trend if one exists. The subgroup size of three also gives the machinist adequate time to accomplish other tasks between taking process measurements.

Data Collection Sheet

Table 9.5. Mount disk hole diameter data collected with $\overline{X}$ and R statistics calculated for $n = 3$.

Subgroup number	1	2	3	4	5	6	7	8	9	10
Sample 1	31	34	34	33	35	34	28	26	33	32
2	34	34	34	32	33	34	27	24	34	32
3	33	33	33	34	33	31	27	25	33	32
$\overline{X}$	32.7	33.7	33.7	33	33.7	33	27.3	25	33.3	32
Range	3	1	1	2	2	3	1	2	1	0

Subgroup number	11	12	13	14	15	16	17	18	19	20
Sample 1	34	33	32	30	33	29	27	33	34	34
2	34	32	34	33	32	30	28	34	32	32
3	34	35	33	34	32	29	24	33	34	34
$\overline{X}$	34	33.3	33	32.3	32.3	29.3	26.3	33.3	33.3	33.3
Range	0	3	2	4	1	1	4	1	2	2

$\overline{X}$ *Chart and Range Chart*

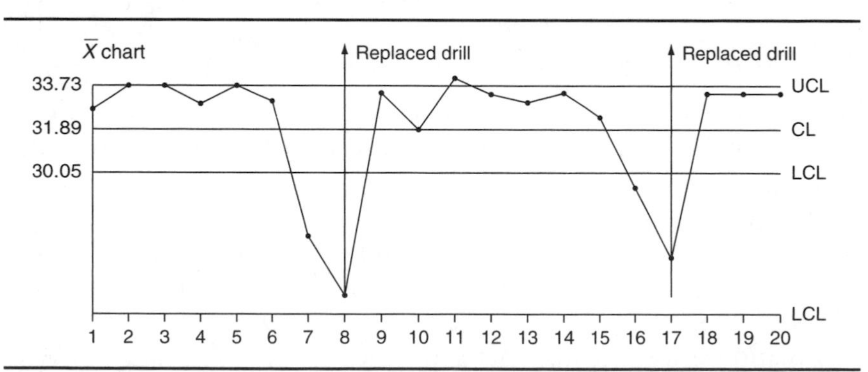

Figure 9.2. Mount disk hole diameter $\overline{X}$ and range control charts.

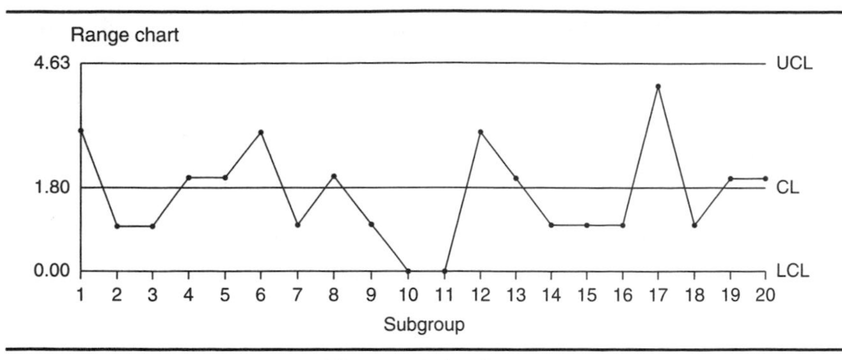

Figure 9.2. Cont.

Calculations for the $\overline{X}$ and Range Chart

$$\overline{\overline{X}} = \frac{\Sigma \overline{X}}{k} = \frac{637.8}{20} = 31.89$$

$$UCL_{\overline{X}} = \overline{\overline{X}} + A_2\overline{R} = 31.89 + 1.023(1.8) = 33.73$$

$$LCL_{\overline{X}} = \overline{\overline{X}} - A_2\overline{R} = 31.89 - 1.023(1.8) = 30.05$$

Calculation 9.1. Calculations for the mount disk hole diameter $\overline{X}$ chart.

$$\overline{R} = \frac{\Sigma R}{k} = \frac{36.0}{20} = 1.80$$

$$UCL_R = D_4\overline{R} = 2.574(1.8) = 4.63$$

$$LCL_R = 0$$

Calculation 9.2. Calculations for the mount disk hole diameter range chart.

Chart Interpretation

Range chart: Shows no out-of-control patterns. There are no shifts, trends, or runs. It appears that the ranges are stable. The 1.80 centerline indicates that the average expected variation within a subgroup of three consecutive drilled holes is 1.80 units, with the largest expected difference to be no more than 4.63 units, as indicated by the UCL_R.

$\overline{X}$ *chart:* Does not indicate a gradual wearing of the drill, but rather a rapid deterioration toward the end of the drill's life. The pattern on the $\overline{X}$ chart clearly shows that the drill maintains a fairly consistent hole size up to the point the tool deteriorates.

When the $\overline{X}$ chart spikes downward, the range chart is not affected. This shows that, when the drill deteriorates and is replaced, it only affects the average of the hole size but not the hole's standard deviation.

Recommendations

- Because the drill deteriorates rapidly toward the end if its life, a standard size drill should be used, but replaced before the rapid deterioration begins.
- To calculate when the tool should be replaced, first collect data to determine the average number of holes a drill cuts before it deteriorates. Next, calculate the standard deviation of this data. The drill should be replaced when the number of holes it has cut equals the average minus three standard deviations.

Estimating the Process Average

The $\overline{X}$ chart is not in control and, therefore, $\overline{\overline{X}}$ is not a reliable estimate of the overall process average. One needs an accurate estimate of the overall process average to calculate process performance ratios.

To accurately estimate the overall process average, the out-of-control plot points on the $\overline{X}$ chart (subgroup numbers 7, 8, 11, 16, and 17) should be removed from the original calculations of $\overline{\overline{X}}$ found in Calculation 9.1.

$$\overline{\overline{X}} = \frac{\Sigma \overline{X}}{k} = \frac{495.9}{15} = 33.06$$

Calculation 9.3. Recalculation of the estimate of the process average (after removing out-of-control plot points).

Estimating σ

Because the range chart is in control, $\overline{R}$ can be used to estimate the value of σ, the process standard deviation (see Calculation 9.4).

$$\hat{\sigma} = \frac{\overline{R}}{d_2} = \frac{1.80}{1.693} = 1.06$$

Calculation 9.4. Estimation of σ using $\overline{R}$.

Calculating Process Capability and Performance Ratios

$$C_p = \frac{USL - LSL}{6\hat{\sigma}} = \frac{35 - 25}{6(1.06)} = \frac{10}{6.36} = 1.57$$

Calculation 9.5. C_p calculation for mount disk diameter.

$$C_{pk_u} = \frac{USL - \overline{\overline{X}}}{3\hat{\sigma}} = \frac{35 - 33.06}{3(1.06)} = \frac{1.94}{3.18} = 0.61$$

Calculation 9.6. $C_{pk\ upper}$ calculation for mount disk diameter.

$$C_{pk_l} = \frac{\overline{\overline{X}} - LSL}{3\hat{\sigma}} = \frac{33.06 - 25}{3(1.06)} = \frac{8.06}{3.18} = 2.53$$

Calculation 9.7. $C_{pk\ lower}$ calculation for mount disk diameter.

Because the C_p value is greater than 1, the process is more than capable of producing almost 100 percent acceptable hole diameters. However, the reported C_{pk} value of 0.61 is less than the C_p value and less than 1. This means the process is not centered and some nonconforming diameter holes are probably being produced.

$\overline{\text{X}}$ *and Range Chart Advantages*

- Separates variation in the average from variation in the standard deviation.
- $\overline{X}$ and range charts are the most widely recognized control charts.
- $\overline{X}$ and range chart principles are used as the foundation for more advanced control charts.

$\overline{\text{X}}$ *and Range Chart Disadvantage*

- Must use a separate $\overline{X}$ and range chart for each characteristic on each part number. This necessitates using multiple $\overline{X}$ and range charts in order to monitor a single part number with several key characteristics.

An Additional Comment About the Case

The machinist was waiting until an out-of-specification hole was detected before changing the drill. (See subgroups 8 and 17 on the control chart.) This is a dangerous practice because all holes between the last "good" subgroup must now be inspected. The money spent chasing down potentially bad parts could exceed the costs of better drill replacement practices.

Chapter 10

$\overline{X}$ and s Chart

Decision Tree Section

Table 10.1. Control chart decision tree.

Number of characteristics or locations on the same chart?	Subgroup size?	Different averages on the same chart?	Different standard deviations on the same chart?	Use this chart
1	≥10	No	No	$\overline{X}$ and s

Description

$\overline{X}$ *Chart*

The $\overline{X}$ chart is used to monitor and detect changes in the average of a single measured characteristic. The plot points represent subgroup averages. The subgroup size, when used in conjunction with an s chart, is typically 10 or more.

s *Chart*

The *s* chart is used to monitor and detect changes in the process standard deviation of a single type of measured characteristic. The plot points represent the calculated sample standard deviation for the 10 or more individual measurements in a subgroup.

Subgroup Assumptions

- Independent measurements
- Constant sample size
- One characteristic
- One unit of measure

Calculating Plot Points

Table 10.2. Plot point formulas for the $\overline{X}$ and *s* chart. Subgroup size is represented by *n*.

Chart	Plot point	Plot point formula
$\overline{X}$	$\overline{X}$	$\frac{\Sigma X}{n}$
s	s (the sample standard deviation)	$\sqrt{\frac{\Sigma(x_i - \overline{X})^2}{n-1}}$

Calculating Centerlines

Table 10.3. Control chart centerline formulas for the $\overline{X}$ and *s* chart. The number of subgroups is represented by *k*.

Chart	Centerline	Centerline formula
$\overline{X}$	$\overline{\overline{X}}$	$\frac{\Sigma \overline{X}}{k}$
s	$\overline{s}$	$\frac{\Sigma s}{k}$

Calculating Control Limits

Table 10.4. Control limit formulas for the $\overline{X}$ and s chart.

Chart	Upper control limit	Lower control limit
$\overline{X}$	$\overline{\overline{X}} + A_3\overline{s}$	$\overline{\overline{X}} - A_3\overline{s}$
s	$B_4\overline{s}$	$B_3\overline{s}$

Example

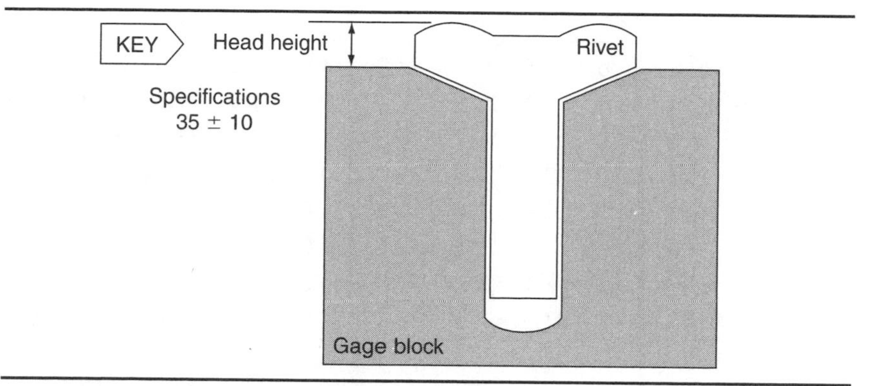

Figure 10.1. Rivet head height measured with the aid of a gage block.

Case Description

Rivet head height is a key characteristic, and it is measured off a gage block. If the height is too low, the installed rivet will recede below the surface. If it is too high, it will protrude. Either case will represent an unacceptable condition requiring rework or reinstallation.

Sampling Strategy

In this manufacturing situation, only one characteristic is being controlled, the production rate is high (thousands per hour), the data collection is quick, and the analysis is completed with the assistance of computer software. For all these reasons, an $\overline{X}$ and s chart was selected for use in helping to control the process.

To determine how often measurements should be taken, the machinist was consulted. It was revealed that adjustments to the equipment must be made about every hour. To capture the effects of these adjustments, data were gathered in sample sizes of 10 every 30 minutes.

Data Collection Sheet

Table 10.5. Data collection sheet for $\overline{X}$ and *s* chart.

Subgroup number	1	2	3	4	5	6	7	8	9	10
Sample 1	36.0	30.6	36.7	37.1	38.2	33.5	34.9	33.0	35.0	33.2
2	38.4	33.3	34.0	35.3	37.1	33.8	37.9	35.6	34.9	32.5
3	37.4	34.8	32.8	30.7	41.3	35.6	35.9	33.8	33.3	37.1
4	36.6	33.0	32.6	33.4	36.2	30.1	37.0	34.0	30.4	36.0
5	37.8	29.7	36.2	30.4	36.5	34.8	36.4	33.1	36.1	36.1
6	39.9	37.2	35.3	36.8	32.6	38.0	32.1	41.7	37.0	28.1
7	36.5	30.5	27.1	36.3	37.9	36.6	29.3	40.1	36.8	35.4
8	31.7	36.0	37.9	35.6	37.5	36.2	40.4	37.8	36.1	30.7
9	31.2	33.0	34.4	38.2	39.4	34.5	35.4	41.0	30.4	32.8
10	31.6	34.1	33.9	34.7	38.1	30.7	37.3	33.3	30.2	41.3
$\overline{X}$	35.7	33.2	34.1	34.9	37.5	34.4	35.7	36.3	34.0	34.3
s	3.11	2.44	2.98	2.63	2.26	2.49	3.10	3.50	2.76	3.69

Subgroup number	11	12	13	14	15	16	17	18	19	20
Sample 1	32.5	34.3	40.0	37.7	39.4	37.8	35.3	34.1	37.1	37.5
2	37.2	35.1	33.1	39.6	38.4	36.3	37.9	35.0	39.0	35.9
3	38.0	36.3	39.0	38.1	38.1	42.8	32.2	33.3	36.9	37.6
4	40.0	26.2	36.6	35.4	31.5	39.0	34.7	30.8	35.6	35.7
5	37.0	39.6	35.9	31.6	35.8	35.0	38.5	38.8	31.7	24.5
6	41.6	37.0	38.3	36.6	38.9	33.1	33.7	37.2	38.3	40.7
7	35.5	29.9	32.8	30.7	39.1	33.7	31.4	36.3	29.8	38.2
8	32.3	31.7	35.7	37.4	33.4	30.2	34.3	32.2	36.9	34.8
9	36.7	35.5	34.6	37.8	37.2	35.0	34.5	36.2	40.7	32.7
10	35.1	33.2	39.7	32.3	31.2	38.7	32.8	35.3	36.4	33.2
$\overline{X}$	36.6	33.9	36.6	35.7	36.3	36.2	34.5	34.9	36.2	35.1
s	2.94	3.84	2.62	3.10	3.17	3.57	2.28	2.39	3.26	4.42

$\overline{X}$ *and* s *Chart*

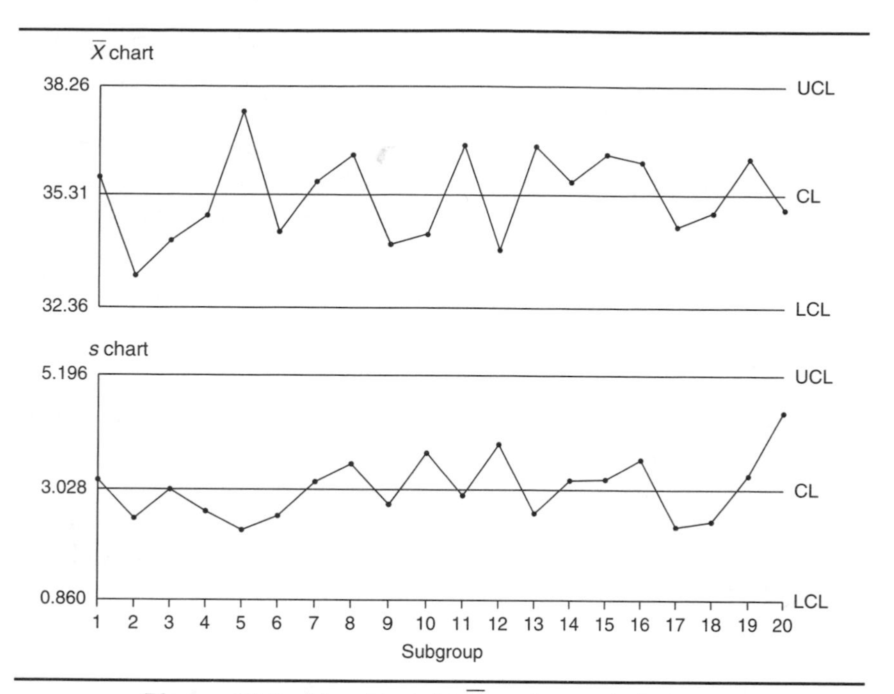

Figure 10.2. Head height $\overline{X}$ and *s* control charts.

Control Limit Calculations

$$\overline{\overline{X}} = \frac{\sum \overline{X}}{k} = \frac{706.1}{20} = 35.31$$

$$UCL_{\overline{X}} = \overline{\overline{X}} + A_3\bar{s} = 35.31 + 0.975(3.028) = 38.26$$

$$LCL_{\overline{X}} = \overline{\overline{X}} - A_3\bar{s} = 35.31 - 0.975(3.028) = 32.36$$

Calculation 10.1. Calculations for rivet head height $\overline{X}$ chart.

$$\bar{s} = \frac{\sum s}{k} = \frac{60.55}{20} = 3.028$$

$$UCL_s = B_4\bar{s} = 1.716(3.028) = 5.196$$

$$LCL_s = B_3\bar{s} = 0.284(3.028) = 0.860$$

Calculation 10.2. Calculations for rivet head height *s* chart.

Chart Interpretation

s *chart:* There are no indications of assignable causes. This means that the variability in the sample standard deviations is consistent from one sampling period to the next.

$\overline{X}$ *chart:* The $\overline{X}$ chart also shows no indications of assignable causes. The subgroup averages are in control. One can expect the rivet head height averages to naturally vary between 32.36 and 38.26 units (the value of the lower and upper control limits, respectively).

Recommendations

- Continue to monitor the rivet head heights for indications of inconsistency and calculate process capability and performance ratios.
- For different rivets, do not assume that head heights will behave in a similar manner. Instead, construct new control charts for different types of rivets.

Estimating the Process Average

Because the $\overline{X}$ chart shows no identifiable assignable causes, the overall process average can be estimated by $\overline{\overline{X}}$.

$$\overline{\overline{X}} = \frac{\Sigma\overline{X}}{k} = \frac{706.1}{20} = 35.31$$

Calculation 10.3. Estimating the overall process average.

Estimating σ

The process standard deviation may be estimated using the s chart's centerline.

$$\hat{\sigma} = \frac{\bar{s}}{c_4} = \frac{3.028}{0.9727} = 3.11$$

Calculation 10.4. Estimating the process standard deviation.

Calculating Process Capability and Performance Ratios

$$C_p = \frac{USL - LSL}{6\hat{\sigma}} = \frac{45 - 25}{6(3.11)} = \frac{20}{18.66} = 1.07$$

Calculation 10.5. C_p calculation for rivet head height.

$$C_{pk_u} = \frac{USL - \overline{\overline{X}}}{3\hat{\sigma}} = \frac{45 - 35.31}{3(3.11)} = \frac{9.69}{9.33} = 1.04$$

Calculation 10.6. $C_{pk\ upper}$ calculation for rivet head height.

$$C_{pk_l} = \frac{\overline{\overline{X}} - LSL}{3\hat{\sigma}} = \frac{35.31 - 25}{3(3.11)} = \frac{10.31}{9.33} = 1.11$$

Calculation 10.7. $C_{pk\ lower}$ calculation for rivet head height.

Because both the C_p and C_{pk} values are greater than 1, it indicates that the process is capable of producing nearly 100 percent acceptable rivet head heights. Because the C_{pk} value is a little bit less than C_p, this means that the process is slightly off-center.

$\overline{X}$ *and* s *Chart Advantages*

- Due to the large subgroup size, the $\overline{X}$ chart is very sensitive to small changes in the overall average.
- When subgroup sizes are large, the sample standard deviation, *s*, is a better indicator of process variation than the range.

$\overline{X}$ *and* s *Chart Disadvantages*

- Large amounts of data need to be gathered before constructing the $\overline{X}$ and *s* chart.
- Because of the large amounts of data, one may need computers and software to efficiently analyze and construct the charts.

Chapter 11

Control Charts for Similar Characteristics— The Target Charts

In this section, three more control charts will be covered. Target control charts will help the SPC practitioner deal with situations where he or she encounters

- Similar characteristics with different dimensions
- Small lot sizes
- High product mix with low production volumes

When these specific conditions are encountered, the SPC practitioner should benefit from using one or a combination of these charts.

- Target *IX*–moving range chart
- Target $\overline{X}$ and range chart™
- Target $\overline{X}$ and *s* chart

Take an example where a machine drills holes for both a 0.25-inch diameter hole and a 0.40-inch diameter hole. For someone familiar with only traditional control charts, this situation would require the use of two separate control charts. Separate traditional control charts would be required for each different size drilled hole because the averages and the scales on the $\overline{X}$ charts would be different. The drill bits and setups would most likely be different, as would be the targets for the hole dimensions of 0.25 inches and 0.40 inches.

However, what is similar is the process itself. The characteristics of interest (hole diameters) are also quite similar. Given these similarities, one may consider using a single target chart to track the consistency of both hole diameters.

The key to using target charts is to identify a target for each characteristic and then mathematically determine how far away the actual measurements fall from the target. In this way, the target becomes a common point from which many similar characteristics with different dimensions can be evaluated. The target value itself is typically identified in one of three different ways.

1. Target defined as print spec nominal (the mid point in the engineering specification).
2. Target defined by the machinist or engineer. Because of manufacturing limitations or some other constraint, one may intentionally center the process at a value that is different than the specification midpoint.
3. Target defined as a value sufficiently far enough away from a maximum only or minimum only (unilateral) specification.

Take, for example, three shafts with three different diameters (see Figure 11.1).

Assume that parts have already been produced and the operator has measured and recorded one diameter measurement from each of parts A, B, and C. The results are shown in Table 11.1.

The deviation from target values are those that are used for plotting on the target *IX* or target $\overline{X}$ charts. They are coded values that represent the distance each measurement falls from its target.

With target charts, the *IX* or $\overline{X}$ chart does not have a traditional vertical axis. Instead, target charts have a zero point representing each characteristic's respective target value.

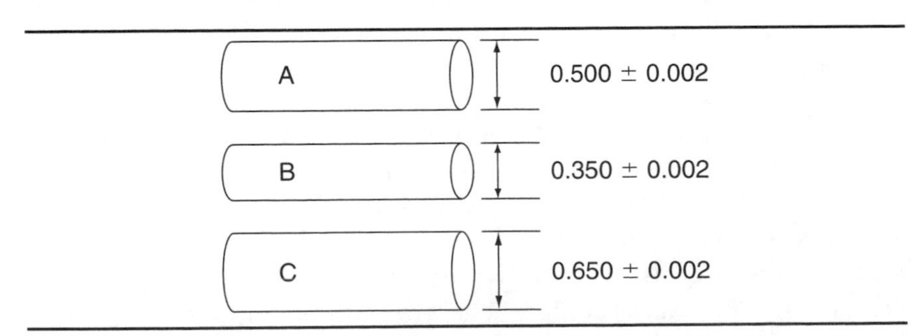

Figure 11.1. Three different size shafts.

Table 11.1. Deviation from target calculations.

Hole	Measurement	Specification	Target	Deviation from target
A	0.503	0.500 ± 0.002	0.500	0.003
B	0.346	0.350 ± 0.002	0.350	−0.004
C	0.651	0.650 ± 0.002	0.650	0.001

If the data being gathered are represented in thousandths of an inch, then the plus and minus zones will be graduated as thousandths of an inch away from the zero point. In addition, characteristics being evaluated on the target charts are separated with vertical lines that partition data of different products or part numbers from one another.

If a plot point on the target $\overline{X}$ chart falls directly on the zero point, it means for that particular subgroup, the average measurement was equal to the target value. A subgroup average that falls in the plus zone (above the zero point) is an indication that the average measurement is larger than the target. Likewise, an average that falls below the zero point represents an average that is smaller than the target. This coding system applies only to the target *IX* or target $\overline{X}$ charts. Range and moving range charts are not affected by this coding method.

Coding the data in this way and using a common vertical axis lets users monitor similar characteristics with different dimensions on the same chart.

Typical Target $\overline{X}$ *and Range*™ *Chart*

Notice that, on the target $\overline{X}$ chart in Figure 11.2, all part A averages fall below the centerline, all part B measurements fall above the centerline, and the part C averages vary around the centerline. This is an example where the patterns are evident of part-specific differences. That is, the averages seem to be different not because of the drilling process itself, but because of some difference between the individual part numbers, their setups, the tooling used in manufacturing the different parts, or some other difference that is specific to the parts themselves.

An example of a process-specific pattern would be a trend that continues across multiple parts (see Figure 11.3).

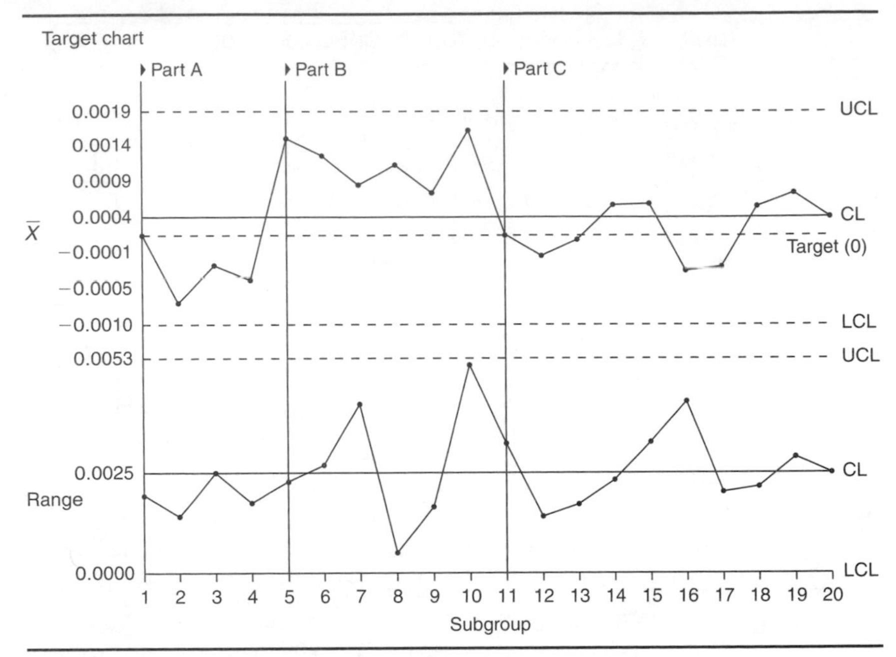

Figure 11.2. Outside diameters from three parts are plotted on the same target $\overline{X}$ and range™ chart.

> **Caution:** When using target charts, subgroups should be made up of measurements of the same characteristic. Never should measurements from different characteristics be included in the same subgroup. Doing so would prove confusing and could result in invalid calculations of plot points and control limits.

When using traditional control charts, each different characteristic, regardless of its similarity to others, requires separate control charts. Given the earlier OD example, one would need to manage three different traditional control charts. Ten similar but different characteristics would require 10 different traditional control charts. When a user has the target chart as his or her SPC tool, most similar characteristics can be monitored on the same chart.

Out-of-Control Situations on Target Charts

When a target *MR*, target *R*, or target *s* chart shows significant differences in variability between part numbers or characteristics, one should not calculate control limits for the target *IX* or target $\overline{X}$ chart.

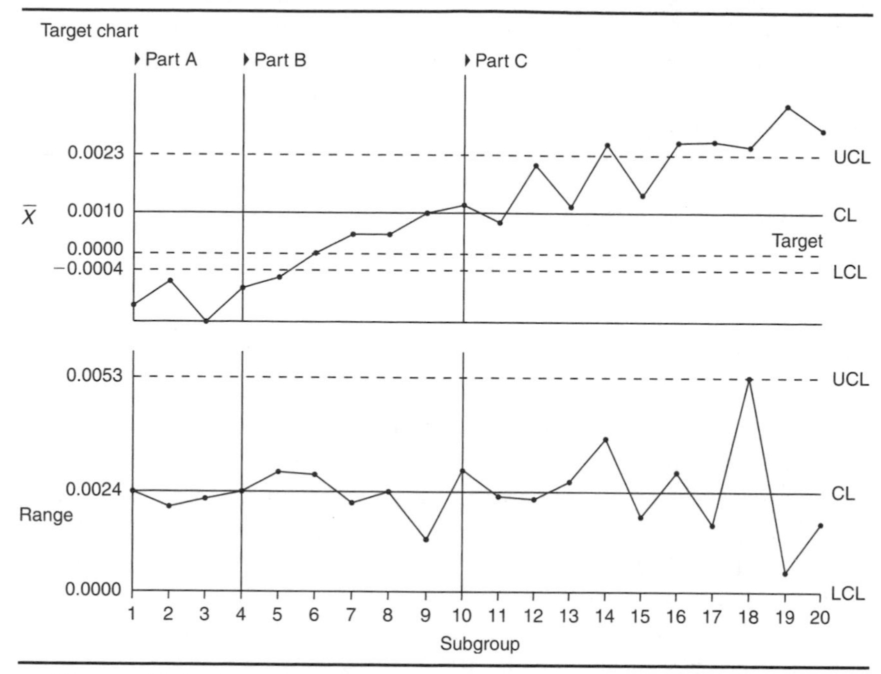

Figure 11.3. A process-specific trend that continues across three parts on the same target $\overline{X}$ and range™ chart.

Recall that control limits from *IX* or $\overline{X}$ charts should be calculated using reliable values of either the $\overline{MR}$, $\overline{R}$, or $\bar{s}$. An inconsistent *MR*, *R*, or *s* chart negates the reliability of their centerlines that are used in calculating control limits for their companion *IX* or $\overline{X}$ charts.

Instead, if one does encounter an inconsistent *MR*, *R*, or *s* chart when used in conjunction with a target *IX* or target $\overline{X}$ chart, the user has these options.

1. Remove the suspect data from the current target chart and place them on a separate target chart made up of similar characteristics with similar levels of variability.
2. Remove the suspect data from the current target chart and place them on their own, dedicated traditional control chart.
3. Revisit the control chart selection tree (Figure 6.1). Notice that the decision tree asks the question, "Are there different standard deviations on the same chart?" If the answer (as in this case) is yes, then the user is directed toward using one of the short run control charts.

There are no special instructions for dealing with assignable causes of variation on target *IX* or target $\overline{X}$ charts. So, when a target *IX* or target $\overline{X}$ chart reveals an out-of-control situation, it should be dealt with like any other *IX* or $\overline{X}$ chart.

Target Chart Benefits

Using target control charts benefits both the machine operator and the organization. For example,

1. The process owner or operator can evaluate process performance across multiple parts on the same chart.
2. Combining similar parts on the same chart allows one to study the consistency of the manufacturing *process* instead of focusing solely on just the consistency of the manufactured *product*.
3. Using the target charts allows the evaluation of patterns and assignable causes that are both process-specific and part-specific.

Chapter 12

Target Individual *X* and Moving Range Chart

Decision Tree Section

Table 12.1. Control chart decision tree.

Number of characteristics or locations on the same chart?	Subgroup size?	Different averages on the same chart?	Different standard deviations on the same chart?	Use this chart
1	1	Yes	No	Target *IX–MR*

Description

Target Individual X *Chart*

The target individual *X* chart, also called a *target IX* or *target X chart,* is used to monitor and detect changes in individual measurements of one characteristic at a time, but allows for several similar characteristics to be plotted on the same chart. This is achieved by coding the data. The coding is the result of subtracting a target value from an actual measurement. Target values are typically, although not always, set at the desired centering of the process, usually specification nominal. Each distinct characteristic on the chart will have its own target value.

These charts are mostly used to monitor similar characteristics with different dimensions from processes where only one measurement represents the process at a given period of time or from processes that produce small numbers of the product or characteristic being evaluated. Examples include measurements from homogeneous batches like chemical concentrations or a dimensional feature such as an OD, with different nominal values from small lot sizes that are manufactured on the same machine.

Moving Range Chart

The moving range chart, also called an *MR chart,* is used to monitor and detect changes in the variation from one individual measurement to the next. The plot points represent the absolute difference between two consecutive individual measurements. The *MR* chart is not affected by the target coding.

Subgroup Assumptions

- Independent measurements
- Constant sample size, $n = 1$
- Similar characteristics
- One unit of measure
- Individual measurements normally distributed
- Similar $\overline{MR}$s among the different plotted characteristics

Calculating Plot Points

Table 12.2. Plot point formulas for the target *IX* and *MR* chart.

Chart	Plot point	Plot point formula
Target individual *X*	Coded *IX*	*IX* – Target
Moving range	*MR*	Absolute difference between two consecutive coded *IX* measurements

Calculating Centerlines

Table 12.3. Control chart centerline formulas for the target *IX* and *MR* chart.

Chart	Centerline	Centerline formula
Target individual *X*	Coded $\overline{IX}$	$\frac{\sum \text{Coded } IX}{k}$
Moving range	$\overline{MR}$	$\frac{\sum MR}{k-1}$

Calculating Control Limits

Table 12.4. Control limit formulas for the target *IX* and *MR* chart.

Chart	Upper control limit	Lower control limit
Target individual *X*	Coded $\overline{IX} + A_2\overline{MR}$	Coded $\overline{IX} - A_2\overline{MR}$
Moving range	$D_4\overline{MR}$	$D_3\overline{MR}$

Example

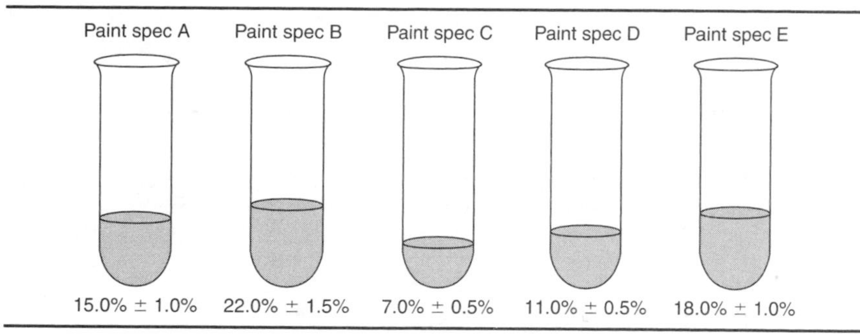

Figure 12.1. Target percent solids from five different paint specifications.

Case Description

Solids content in paint is a key characteristic. To obtain a measure of solids content, a paint sample of known weight is taken from a mixing tank—one sample per paint batch. The sample is baked in an oven until only solids remain. The remaining solids are weighed and a percent solids is calculated. In this example, a mixing tank is used to produce five different types of paint: A, B, C, D, and E. Each paint type requires a different percent solids content. Long production runs rarely occur with any one paint. The production manager is monitoring the solids content from all five paints on the same SPC chart.

Sampling Strategy

A target *IX–MR* chart is used to monitor this process because

1. Only one characteristic is being controlled (solids content).
2. One measurement is representative of each batch.
3. The user prefers to construct a single chart to track multiple paint specs.

Data Collection Sheet

Table 12.5. Data collection sheet for constructing target individual *X* and moving range chart.

Subgroup	1	2	3	4	5	6	7	8	9	10
Batch	D	E	E	D	A	E	E	A	A	D
IX	10.7	18.1	18.1	10.7	14.8	17.9	17.8	14.9	14.9	11.0
Target	11.0	18.0	18.0	11.0	15.0	18.0	18.0	15.0	15.0	11.0
IX – Target	−0.3	0.1	0.1	−0.3	−0.2	−0.1	−0.2	−0.1	−0.1	0.0
MR	—	0.4	0.0	0.4	0.1	0.1	0.1	0.1	0.0	0.1
Subgroup	11	12	13	14	15	16	17	18	19	20
Batch	D	A	E	C	B	A	A	B	C	C
IX	10.9	14.7	17.6	7.5	22.0	15.3	15.0	22.2	7.1	7.4
Target	11.0	15.0	18.0	7.0	22.0	15.0	15.0	22.0	7.0	7.0
IX – Target	−0.1	−0.3	−0.4	0.5	0.0	0.3	0.0	0.2	0.1	0.4
MR	0.1	0.2	0.1	0.9	0.5	0.3	0.3	0.2	0.1	0.3

Target Individual X *and Moving Range Chart*

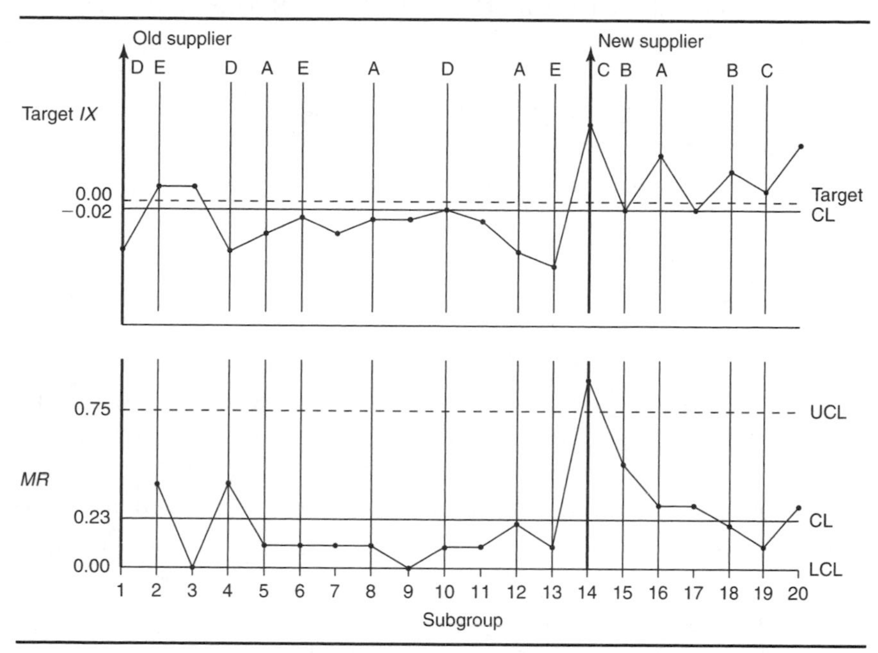

Figure 12.2. Percent solids target *IX* and *MR* chart.

Calculations for the MR *Chart*

$$\overline{MR} = \frac{\sum MR}{k-1} = \frac{4.30}{19} = 0.23$$

$$\mathrm{UCL}_{MR} = D_4\overline{MR} = 3.267(0.23) = 0.75$$

$$\mathrm{LCL}_{MR} = 0$$

Calculation 12.1. Calculations for *MR* chart.

MR *Chart Interpretation and Recalculation*

An upward spike occurs on the *MR* chart when the new supplier's products begin to be used. Because the *MR* chart is out of control, this means that the value of $\overline{MR}$ is unreliable and cannot be used to calculate control limits for the target *IX* chart. This is why no control limits were placed on the target *IX* chart in Figure 12.2.

After removing the out-of-control plot point (subgroup number 14) from the *MR* chart, the $\overline{MR}$ was recalculated using the remaining 18 *MR* values (see Calculation 12.2).

$$\overline{MR} = \frac{\sum MR}{k-1} = \frac{3.40}{18} = 0.19$$

$$\text{UCL}_{MR} = D_4\overline{MR} = 3.267(0.19) = 0.62$$

$$\text{LCL}_{MR} = 0$$

Calculation 12.2. Revised *MR* chart calculations after removing subgroup number 14.

Note that all of the remaining moving range values fall within the new *MR* chart control limits (see Figure 12.3). There appears to be no indication of assignable causes of variation. Given this situation, it is now appropriate to complete the control chart calculations for the target *IX* chart.

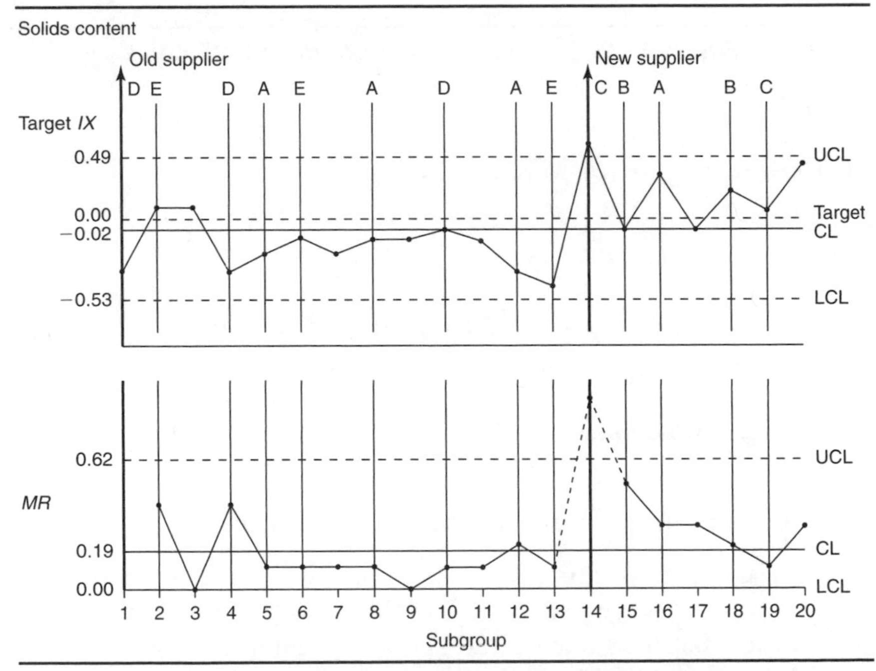

Figure 12.3. Target *IX* and *MR* chart with revised control limits. Subgroup number 14 has been removed from calculations for the *MR* chart.

Calculations for the Target IX *Chart*

$$\text{Coded } \overline{IX} = \frac{\sum \text{Coded } IX}{k} = \frac{-.40}{20} = -0.02$$

$$\text{UCL}_{IX} = \text{Coded } \overline{IX} + A_2\overline{MR} = -0.02 + 2.660(0.19) = 0.49$$

$$\text{LCL}_{IX} = \text{Coded } \overline{IX} - A_2\overline{MR} = -0.02 - 2.660(0.19) = -0.53$$

Calculation 12.3. Calculations for the percent solids target *IX* chart.

Target IX *Chart Interpretation*

It appears that, after the supplier change, the percent solids contents increased across paints A, B, and C. The run above the centerline between plot points 14 and 20 was determined to be the result of changing the supplier. The run below the centerline between points 4 and 9 is, in part, due to the upward shift in the centerline between points 14 and 20.

> **Note:** When analyzing target charts, also look for patterns unique to each characteristic represented on the chart. For example, look to see if all of paint A plot points were above or below the centerline or trending upward or downward. In this example, all paint B plot points are above the centerline, but there are only two plot points. This does not qualify as an assignable cause. However, if eight or more plot points from the same paint were above the centerline, it would indicate an out-of-control condition unique to that paint. This would be true regardless of how many different paints were manufactured between those points.

Recommendation

Supplier changes should not be introduced into the line without first knowing how the change will affect the producibility and/or the finished product. If the effects are known in advance, prior adjustments can possibly be made without affecting the production line. In many cases, the costs associated with changing suppliers exceed the benefits of a lower price.

Estimating the Process Average

The coded $\overline{IX}$ on the control chart (−0.02 percent) has been upwardly influenced because of the supplier change assignable cause. Because of the presence of an assignable cause, the overall average of −0.02 percent is not a reliable estimate of the centering of the process.

To accurately estimate the overall process average, we will evaluate only the data from the old supplier (the first 13 subgroups). This data by itself proved to be in control on a separate target *IX* and *MR* chart (not shown here).

$$\text{Coded } \overline{IX} = \frac{\sum \text{Coded } IX}{k} = \frac{-1.9}{13} = -0.15$$

Calculation 12.4. Estimate of the process average based upon old supplier data (first 13 subgroups).

The coded $\overline{IX}$ from Calculation 12.4 shows that, on average, each old supplier batch of paint is approximately 0.15 percent below targets. If enough data were gathered from the new supplier data, it might be interesting to evaluate the old supplier's coded $\overline{IX}$ with the new supplier's $\overline{IX}$.

Estimating σ

The moving range chart for the first 13 subgroups (not shown) proved to be in control. The calculation for $\overline{MR}$ is shown in Calculation 12.5.

$$\overline{MR} = \frac{\sum MR}{k-1} = \frac{1.70}{12} = 0.14$$

Calculation 12.5. Average moving range calculation from first 13 subgroups.

$$\hat{\sigma} = \frac{\overline{MR}}{d_2} = \frac{0.14}{1.128} = 0.12$$

Calculation 12.6. Estimating sigma using $\overline{MR}$ from Calculation 12.5.

Note that the first 13 subgroups represent only old supplier data. Therefore, the $\hat{\sigma}$ found in Calculation 12.6 can be thought of as the estimated standard deviation for the old supplier. Notice, though, that the first 13 subgroups also are representative of process performance from paint specs A, D, and E. No data representing paint specs B or C are found. Therefore, paint specs A, D, and E will be used in calculating C_p and C_{pk} values. There will be no calculation of C_p or C_{pk} values for paint specs B or C.

Calculating Process Capability and Performance Ratios

Capability ratios will be calculated for each paint specification found in the first 13 subgroups. Because the *MR* chart is in control, the same $\hat{\sigma}$ may be used in calculating process capability and performance ratios for paint specifications A, D, and E. The C_p calculation for paint specification *A* (assuming the old supplier's materials are used) is found in Calculation 12.7.

$$C_{pA} = \frac{\text{USL}_A - \text{LSL}_A}{6\hat{\sigma}} = \frac{16.0 - 14.0}{6(0.12)} = \frac{2.0}{0.72} = 2.78$$

Calculation 12.7. Process capability ratio for paint spec A using old supplier data.

In order to calculate $C_{pkA,}$ the process average must first be estimated for paint spec A. The estimate of the paint spec A process average is given in Calculation 12.8.

$$\overline{IX}_A = \frac{14.8 + 14.9 + 14.9 + 14.7}{4} = \frac{59.3}{4} = 14.83$$

Calculation 12.8. Estimate of the process average for paint spec A.

$$C_{pk_uA} = \frac{\text{USL}_A - \overline{IX}_A}{3\hat{\sigma}} = \frac{16.0 - 14.83}{3(0.12)} = \frac{1.17}{0.36} = 3.25$$

Calculation 12.9. $C_{pk\ upper}$ calculation for paint spec A.

$$C_{pk_lA} = \frac{\overline{IX}_A - \text{LSL}_A}{3\hat{\sigma}} = \frac{14.83 - 14.0}{3(0.12)} = \frac{0.83}{0.36} = 2.31$$

Calculation 12.10. $C_{pk\ lower}$ calculation for paint spec A.

Because the C_p value is greater than 1, the process is more than capable of producing almost 100 percent acceptable output. Because the C_{pk} value is smaller than the C_p value, it means that the process is a little off center, but because the C_{pk} value is larger than 1, the process is performing to specifications.

The C_p and C_{pk} ratios for paint specs D and E can be found in Table 12.6.

> **Note:** To ensure reliable estimates of σ and the process average, one needs about 20 data points. Therefore, the calculations on these pages and those in Table 12.6 are used for illustration purposes only.

Target IX–MR *Chart Advantages*

- Multiple parts, specifications, or characteristics can be plotted on the same chart (provided they all have similar variability as exhibited by an in-control *MR* chart).
- C_p and C_{pk} can be calculated for each characteristic on the chart.
- Statistical control can be assessed for both the process and each unique part and/or characteristic on the chart.

Target IX–MR *Chart Disadvantages*

- When interpreting the target *IX* chart, both the zero line and the coded $\overline{X}$ must be taken into account.
- The moving range plot points are dependent on the individual *X* plot points. In other words, changes in the *MR* chart are directly related to changes from one individual measurement to the next.
- Variation in the individual measurements could be caused by a shift in the average or the inherent standard deviation of the process; however, the *IX–MR* charts cannot efficiently separate the effects of the two.
- Reliable control limits require the distribution of the individual measurements to be approximately normal.
- The target *IX–MR* chart is not as sensitive to changes in the process average or standard deviation as would be a target $\overline{X}$ and range chart™.

Additional Comments About the Case

C_p and C_{pk} for paint specifications D and E are shown in Table 12.6.

Table 12.6. C_p and C_{pk} values for paint specifications D and E.

Paint spec D	Paint spec E
$\overline{IX}_D = 10.83$	$\overline{IX}_E = 17.90$
$C_{pD} = 1.39$	$C_{pE} = 2.78$
$C_{pk_uD} = 1.86$	$C_{pk_uE} = 3.06$
$C_{pk_lD} = 0.92$	$C_{pk_lE} = 2.50$

Chapter 13

Target $\overline{X}$ and Range™ Chart

Decision Tree Section

Table 13.1. Control chart decision tree.

Number of characteristics or locations on the same chart?	Subgroup size?	Different averages on the same chart?	Different standard deviations on the same chart?	Use this chart
1	>1 <10	Yes	No	Target $\overline{X}$ and range™

Description

Target $\overline{X}$ Chart

The target $\overline{X}$ chart is used to monitor and detect changes in the average of a single type of measured characteristic—regardless of the part number. However, the part number can only be changed between subgroups. Target values change when the respective part number changes. The plot points represent a subgroup's average deviation from its target value. Target values are set at the desired centering of the process, which is typically, although not always, specification nominal (the middle of two-sided specs). Subgroup sizes typically range from two to nine but practitioners generally use subgroup sizes of three or five.

Range Chart

The range chart, also called an *R chart,* is used to monitor and detect changes in the standard deviation of a single type of measured characteristic. The plot points represent subgroup ranges. Ranges are not the same as standard deviation, but when subgroup sizes are small—less than 10—ranges can be used to estimate within subgroup standard deviation. The range chart is not affected by the "deviation from target" data coding that takes place with the target $\overline{X}$ chart.

Subgroup Assumptions

- Independent measurements
- Constant sample size
- Similar characteristics
- One unit of measure
- Similar range values among the different characteristics on the chart.

Calculating Plot Points

Table 13.2. Plot point formulas for the target $\overline{X}$ and range™ chart.

Chart	Plot point	Plot point formula
Target $\overline{X}$	Coded $\overline{X}$	$\overline{X}$ – target
Range	R	Largest X – Smallest X

Calculating Centerlines

Table 13.3. Control chart centerline formulas for the target $\overline{X}$ and range™ chart, where k is the number of subgroups.

Chart	Centerline	Centerline formula
Target $\overline{X}$	Coded $\overline{\overline{X}}$	$\dfrac{\Sigma \text{Coded } \overline{X}}{k}$
Range	$\overline{R}$	$\dfrac{\Sigma R}{k}$

Calculating Control Limits

Table 13.4. Control limit formulas for the target $\overline{X}$ and range™ chart.

Chart	Upper control limit	Lower control limit
Target $\overline{X}$	Coded $\overline{\overline{X}} + A_2\overline{R}$	Coded $\overline{\overline{X}} - A_2\overline{R}$
Range	$D_4\overline{R}$	$D_3\overline{R}$

Example

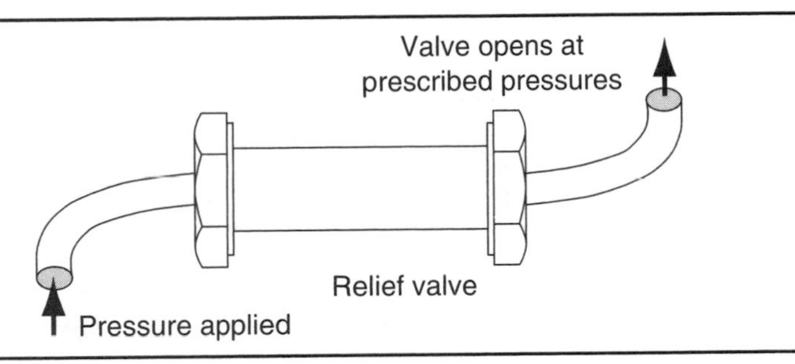

Figure 13.1. Relief valve with adjustable cracking pressure capabilities.

Case Description

Cracking pressure, the pressure at which the relief valve opens, is a key characteristic. The valve can be adjusted during assembly to crack at different pressures. Each customer has his or her own crack pressure requirements.

Sampling Strategy

Cracking pressure is the only characteristic, but the requirements change with each order (see Table 13.5). Because the production vol-

Table 13.5. Crack pressure requirements for three valve customers.

Valve customer	Target crack pressure	Specifications
A	220 psi	±3%
B	270 psi	±2%
C	180 psi	±4%

ume is steady and the standard deviation is expected to be consistent across all cracking pressure settings, a target $\overline{X}$ and range™ chart is used to monitor the process. Valves are 100 percent tested, but for charting purposes, the test results from three out of every 30 valves are used for analysis on control charts.

Data Collection Sheet

Table 13.6. Data collection sheet for relief valves.

Subgroup number	Customer A								
	1	2	3	4	5	6	7	8	9
Sample 1	219	220	219	220	220	217	219	219	220
2	219	217	220	217	217	219	218	222	220
3	223	219	223	220	222	221	221	222	218
$\overline{X}$	220.3	218.7	220.7	219.0	219.7	219.0	219.3	221.0	219.3
Target	220	220	220	220	220	220	220	220	220
$\overline{X}$ – Target	0.3	−1.3	0.7	−1.0	−0.3	−1.0	−0.7	1.0	−0.7
Range	4	3	4	3	5	4	3	3	2

Subgroup number	Customer B					
	10	11	12	13	14	15
Sample 1	271	269	270	272	272	273
2	272	274	268	271	272	270
3	270	270	273	271	270	267
$\overline{X}$	271.0	271.0	270.3	271.3	271.3	270.0
Target	270	270	270	270	270	270
$\overline{X}$ – Target	1.0	1.0	0.3	1.3	1.3	0.0
Range	2	5	5	1	2	6

Subgroup number	Customer C				
	16	17	18	19	20
Sample 1	175	177	174	176	175
2	178	178	177	178	178
3	175	176	173	173	175
$\overline{X}$	176.0	177.0	174.7	175.7	176.0
Target	180	180	180	180	180
$\overline{X}$ – Target	−4.0	−3.0	−5.3	−4.3	−4.0
Range	3	2	4	5	3

Target $\overline{\text{X}}$ *and Range™ Chart*

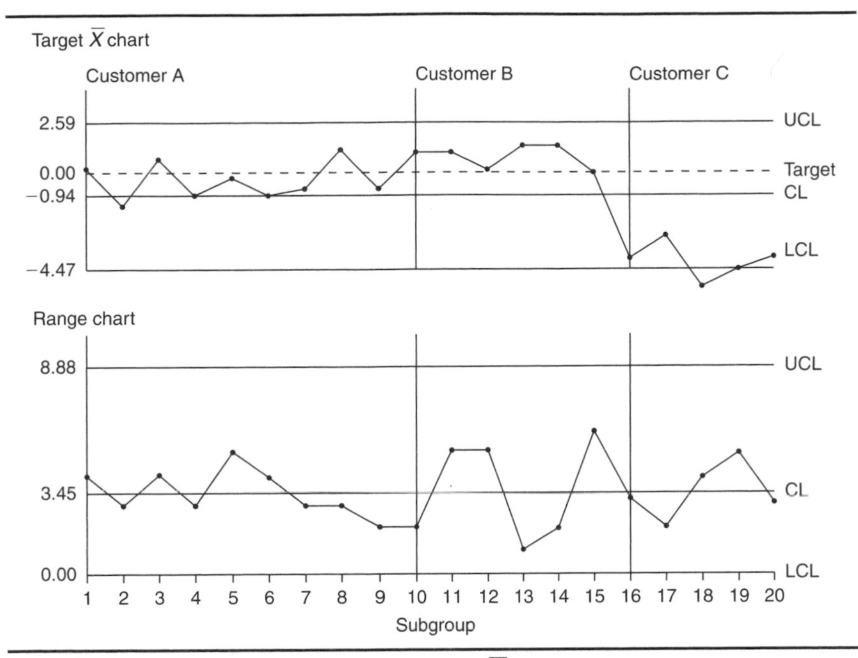

Figure 13.2. Crack pressure target $\overline{X}$ and range™ control chart.

Control Limit Calculations

$$\text{Coded } \overline{\overline{X}} = \frac{\sum \text{Coded } \overline{X}}{k} = \frac{-18.7}{20} = -0.94$$

$$\text{UCL}_{\overline{X}} = \text{Coded } \overline{\overline{X}} + A_2\overline{R} = -0.94 + 1.023(3.45) = 2.59$$

$$\text{LCL}_{\overline{X}} = \text{Coded } \overline{\overline{X}} - A_2\overline{R} = -0.94 - 1.023(3.45) = -4.47$$

Calculation 13.1. Calculations for the crack pressure target $\overline{X}$ chart.

$$\overline{R} = \frac{\sum R}{k} = \frac{69.0}{20} = 3.45$$

$$\text{UCL}_R = D_4\overline{R} = 2.574(3.45) = 8.88$$

$$\text{LCL}_R = 0$$

Calculation 13.2. Calculations for the crack pressure range chart.

Chart Interpretation

Range chart: No out-of-control plot points. There are no shifts, trends, or runs. It appears that the ranges are stable. This normal pattern supports the assumption that the process standard deviation is not affected when the valves are adjusted to different cracking pressures.

Target $\overline{X}$ chart: Plot point comparisons to both the coded $\overline{\overline{X}}$ and the zero line must be made. Relative to the coded $\overline{\overline{X}}$ (-0.94) none of the jobs is centered; this is caused mainly by customer C's job being run well below its target of 180 psi. These plot points are pulling down the entire average, thus causing there to appear significantly long runs of plot points above the coded $\overline{\overline{X}}$.

Relative to the zero line, the valve for customer A is centered on target, valves for customer B are a little on the high side of the target, and customer C's valves are running consistently low.

Recommendations

If a characteristic is not centered on its target, either the process needs to be adjusted or the target needs to be changed.

Assuming the targets are desired values,

- Customer A valves are centered on target; no adjustment needs to be made.
- Customer B valves are a little on the high side. The benefit of centering the crack pressure on its target may not be worth the effort required if the C_p and C_{pk} values are high (greater than 1.3).
- Customer C valves need to be adjusted about 5 psi higher. However, before changing the process, people attending to the process should verify the off-target values are not caused by a faulty measurement system.

Estimating the Process Average

The average difference from target is not the same for all three valve adjustments. So calculations for $\overline{\overline{X}}$ need to be done separately for each of the three customer requirements. The following example focuses on customer A valves.

$$\overline{\overline{X}}_A = \frac{\Sigma \overline{X}_A}{k} = \frac{1977.0}{9} = 219.67$$

Calculation 13.3. Calculation for customer A's average cracking pressure.

Note: To ensure reliable estimates, k should be about 20. In this example k is only nine. Therefore, the calculations on these pages and in the additional comments section are used only for illustration purposes.

Estimating σ

Because the range chart is in control across all three customer requirements, the estimate of sigma for all valves may be based upon the range chart's centerline (see Calculation 13.4). If the range chart were not in control, separate, reliable $\overline{R}$ values would need to be calculated for each of the customer requirements.

$$\hat{\sigma} = \frac{\overline{R}}{d_2} = \frac{3.45}{1.693} = 2.04$$

Calculation 13.4. Estimating sigma using $\overline{R}$.

Calculating Process Capability and Performance Ratios

Because the R chart is in control, the same $\hat{\sigma}$ may be used for separately calculating all process capability and performance ratios for the cracking pressures. Following are the C_p and C_{pk} calculations for customer A valves.

$$C_{pA} = \frac{USL_A - LSL_A}{6\hat{\sigma}} = \frac{226.6 - 213.4}{6(2.04)} = \frac{13.2}{12.24} = 1.08$$

Calculation 13.5. C_p calculation for customer A valves.

$$C_{pk_u A} = \frac{USL_A - \overline{\overline{X}}_A}{3\hat{\sigma}} = \frac{226.6 - 219.67}{3(2.04)} = \frac{6.93}{6.12} = 1.13$$

Calculation 13.6. $C_{pk\ upper}$ calculation for customer A valves.

$$C_{pk_l A} = \frac{\overline{\overline{X}}_A - LSL_A}{3\hat{\sigma}} = \frac{219.67 - 213.4}{3(2.04)} = \frac{6.27}{6.12} = 1.02$$

Calculation 13.7. $C_{pk\ lower}$ calculation for customer A valves.

Target $\overline{\text{X}}$ and Range™ Chart Advantages

- Multiple parts, specifications, or characteristics can be plotted on the same chart (provided they all exhibit similar variability).
- Data from gages that are zeroed out on their target values can be plotted directly on the target $\overline{X}$ without further data coding or transformation.
- Statistical control can be assessed for both the process and each unique part and/or characteristic being made.

Target $\overline{\text{X}}$ and Range™ Chart Disadvantages

- Control limits are valid only when the $\overline{R}$s from each part on the chart are similar. When they are not similar, the suspect part(s) must be monitored on a separate chart, or the data must be collectively evaluated on a short run chart.
- When interpreting the target $\overline{X}$ chart, both the zero line and the coded $\overline{\overline{X}}$ must be taken into account. This accounts for some added complexity when interpreting the chart.

Additional Comments About the Case

- The process capability and performance ratio calculations for the cracking pressure are shown in Table 13.7.
- When valves A, B, or C are run again, the new data can be combined with prior data.

Table 13.7. C_p and C_{pk} calculations for valves B and C.

Valve B	Valve C
$\overline{\overline{X}}_B = 270.82$	$\overline{\overline{X}}_C = 175.88$
$C_{pB} = 0.88$	$C_{pC} = 1.18$
$C_{pk_uB} = 0.75$	$C_{pk_uC} = 1.85$
$C_{pk_lB} = 1.02$	$C_{pk_lC} = 0.50$

Chapter 14

Target $\overline{X}$ and s Chart

Decision Tree Section

Table 14.1. Control chart decision tree.

Number of characteristics or locations on the same chart?	Subgroup size?	Different averages on the same chart?	Different standard deviations on the same chart?	Use this chart
1	≥10	Yes	No	Target $\overline{X}$ and s

Description

Target $\overline{X}$ *Chart*

The target $\overline{X}$ chart is used to monitor and detect changes in the average of a single type of measured characteristic—regardless of the part number. However, the part number can only be changed between subgroups. Target values change when the respective part number changes. The plot points represent a subgroup's average deviation from its target value. Target values are set at the desired centering of the process, which is typically, although not always, specification nominal (the middle of two-sided specs). The subgroup size, when used in conjunction with an *s* chart, is typically 10 or more.

s *Chart*

The *s* chart is used to monitor and detect changes in the standard deviation of a single type of measured characteristic. The plot points are the calculated sample standard deviation for the 10 or more individual measurements in a subgroup. The *s* chart is not affected by the "deviation from target" data coding that takes place with the target $\overline{X}$ chart.

Subgroup Assumptions

- Independent measurements
- Constant sample size
- Similar characteristics
- One unit of measure
- Similar sample standard deviation values among the different characteristics

Calculating Plot Points

Table 14.2. Plot point formulas for the target $\overline{X}$ and *s* chart.

Chart	Plot point	Plot point formula
Target $\overline{X}$	Coded $\overline{X}$	$\overline{X}$ − Target
s	s (sample standard deviation)	$\sqrt{\dfrac{\Sigma(x_i - \overline{X})^2}{n-1}}$

Calculating Centerlines

Table 14.3. Control chart centerline formulas for the target $\overline{X}$ and *s* chart, where *k* is the number of subgroups.

Chart	Centerline	Centerline formula
Target $\overline{X}$	Coded $\overline{\overline{X}}$	$\dfrac{\Sigma \text{Coded } \overline{X}}{k}$
s	$\overline{s}$	$\dfrac{\Sigma s}{k}$

Calculating Control Limits

Table 14.4. Control limit formulas for the target $\overline{X}$ and s chart.

Chart	Upper control limit	Lower control limit
Target $\overline{X}$	Coded $\overline{\overline{X}} + A_3\overline{s}$	Coded $\overline{\overline{X}} - A_3\overline{s}$
s	$B_4\overline{s}$	$B_3\overline{s}$

Example

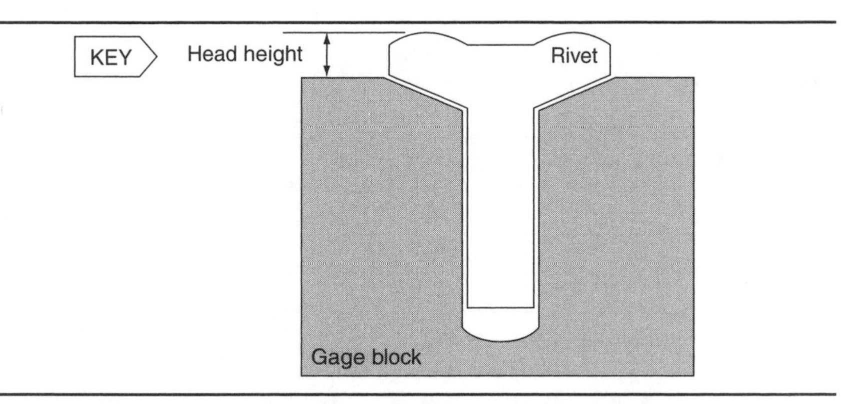

Figure 14.1. Rivet head height is a key characteristic. The measurement is taken with the aid of a gage block.

Case Description

Rivet head height is a key characteristic. The height is measured off a gage block. If the height is too low, the installed rivet will recede below the surface. If it is too high, it will protrude. Either case requires rework and is unacceptable. Three different types of rivets are manufactured, each with different target head heights and tolerances.

Table 14.5. Target head heights and specifications.

Rivet type	Target head height	Specifications
A	35	±10
B	16.5	±5
C	75	±15

Sampling Strategy

Several rivet types are to be plotted on the same chart, but because only one characteristic, head height, is to be controlled, use of a target chart would be appropriate. The production volume is extremely high (thousands per hour), the data collection is quick, and the analysis is being done with the assistance of computer software. For all these reasons, a target $\overline{X}$ and s chart is selected.

To determine how often measurements should be taken, a header mechanic is surveyed. It is revealed that adjustments to the equipment affecting head height are made about every hour. To capture the effects of these adjustments, samples of 10 are taken every 10 minutes.

Data Collection Sheet

Table 14.6. Data collection sheet for three different rivet head heights.

Rivet type				A			
Subgroup number	1	2	3	4	5	6	7
Sample 1	35.9	35.5	34.1	34.1	35.8	37.1	35.1
2	34.3	35.0	37.4	35.2	35.5	35.4	35.3
3	32.6	35.2	34.6	30.7	35.2	37.0	33.0
4	34.2	36.2	35.5	36.9	34.3	34.2	36.2
5	35.9	35.0	34.0	37.9	32.8	34.4	33.4
6	35.7	35.2	34.8	33.7	35.2	36.6	35.3
7	35.0	35.5	36.2	32.6	36.5	33.6	35.4
8	37.0	35.3	34.5	34.1	36.4	35.9	37.6
9	34.7	34.7	36.2	34.7	34.7	37.9	34.0
10	35.1	35.5	35.0	36.1	31.1	34.4	35.4
$\overline{X}$	35.0	35.3	35.2	34.6	34.8	35.7	35.1
Target	35.0	35.0	35.0	35.0	35.0	35.0	35.0
$\overline{X}$ − Target	0.0	0.3	0.2	−0.4	−0.3	0.7	0.1
s	1.21	0.41	1.09	2.09	1.67	1.47	1.34

Table 14.6. Cont.

Rivet type	B				
Subgroup number	8	9	10	11	12
Sample 1	19.0	17.5	18.0	13.7	14.9
2	15.5	16.6	18.0	17.7	16.0
3	16.8	16.4	17.9	13.9	16.4
4	16.6	18.4	14.5	16.8	14.9
5	15.7	16.2	14.7	15.1	15.5
6	16.2	14.3	14.4	15.2	15.3
7	17.4	17.1	18.5	18.7	17.4
8	15.6	17.9	18.0	17.3	15.9
9	19.2	16.4	16.6	17.0	14.8
10	16.7	14.4	17.5	17.3	16.1
$\overline{X}$	16.9	16.5	16.8	16.3	15.7
Target	16.5	16.5	16.5	16.5	16.5
$\overline{X}$ – Target	0.4	0.0	0.3	−0.2	−0.8
s	1.32	1.35	1.65	1.69	0.81

Rivet type	C							
Subgroup number	13	14	15	16	17	18	19	20
Sample 1	72.8	76.5	72.9	73.3	75.0	76.4	74.5	75.6
2	76.3	72.7	73.9	77.5	74.1	73.8	75.1	74.4
3	74.7	75.2	74.5	74.6	75.6	74.2	74.9	74.5
4	76.6	75.2	76.6	75.5	75.9	72.0	74.4	76.4
5	75.0	75.2	75.1	74.4	75.1	74.4	75.3	75.6
6	75.5	75.6	75.2	75.6	75.4	73.7	76.4	73.5
7	76.7	76.9	75.5	71.9	75.2	76.8	76.6	76.6
8	76.1	73.7	74.0	76.8	75.5	76.8	77.0	76.0
9	76.9	74.1	75.0	74.3	72.3	73.0	75.9	73.3
10	77.2	78.5	76.7	77.1	75.3	72.9	75.1	74.9
$\overline{X}$	75.8	75.4	74.9	75.1	74.9	74.4	75.5	75.1
Target	75.0	75.0	75.0	75.0	75.0	75.0	75.0	75.0
$\overline{X}$ – Target	0.8	0.4	−0.1	0.1	−0.1	−0.6	0.5	0.1
s	1.33	1.67	1.18	1.76	1.04	1.71	0.90	1.15

Target $\overline{X}$ *and* s *Chart*

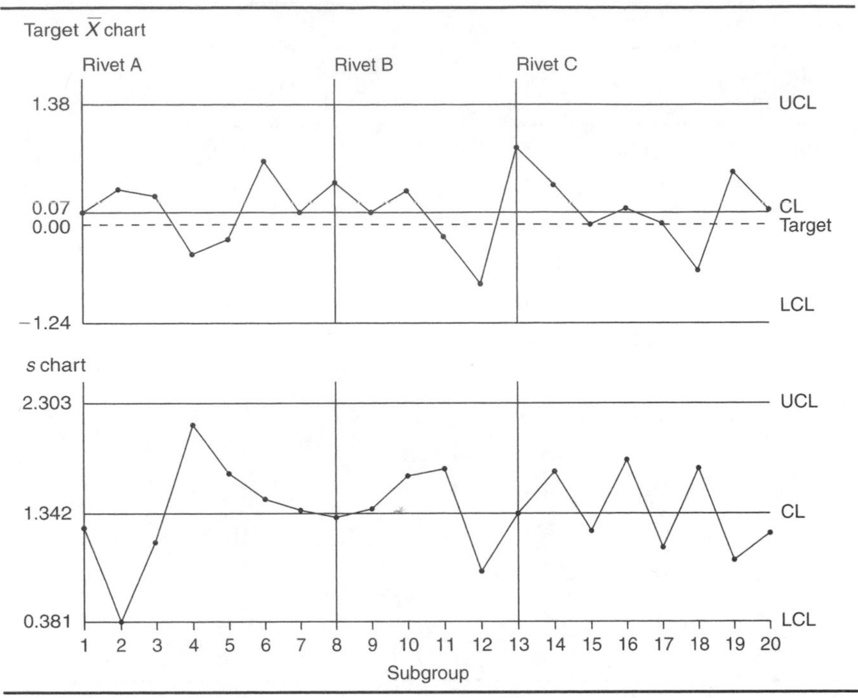

Figure 14.2. Head height target $\overline{X}$ and s control chart.

Control Limit Calculations

$$\text{Coded } \overline{\overline{X}} = \frac{\sum \text{Coded } \overline{X}}{k} = \frac{1.40}{20} = 0.07$$

$$\text{UCL}_{\overline{X}} = \text{Coded } \overline{\overline{X}} + A_3\bar{s} = 0.07 + 0.975(1.342) = 1.38$$

$$\text{LCL}_{\overline{X}} = \text{Coded } \overline{\overline{X}} - A_3\bar{s} = 0.07 - 0.975(1.342) = -1.24$$

Calculation 14.1. Calculations for target $\overline{X}$ chart.

$$\bar{s} = \frac{\sum s}{k} = \frac{26.84}{20} = 1.342$$

$$\text{UCL}_s = B_4\bar{s} = 1.716(1.342) = 2.303$$

$$\text{LCL}_s = B_3\bar{s} = 0.284(1.342) = 0.381$$

Calculation 14.2. Calculations for s chart.

Chart Interpretation

s *chart:* The chart is in control. This shows that the sample standard deviations of head heights for all three rivet types are similar.

Target $\overline{X}$ *chart:* This chart is also in control. There are no indications of assignable causes. This means that the difference between the average head heights of all three rivet types and their respective targets is about the same.

Recommendations

- Based on the target $\overline{X}$ chart, the process is running very close to target regardless of rivet type. This is a situation where the process should not be adjusted.
- Even though the standard deviations are similar for all three rivet types, one will still need to calculate separate C_p and C_{pk} ratios. This is necessary because the engineering tolerances are different for each rivet type.

Estimating the Process Average

Because the target $\overline{X}$ chart is in control, the process average for all rivet types can be estimated using the coded $\overline{\overline{X}}$.

$$\text{Coded } \overline{\overline{X}} = \frac{\sum \text{Coded } \overline{X}}{k} = \frac{1.40}{20} = 0.07$$

Calculation 14.3. Estimate for the coded overall process average rivet head height (to be used in C_{pk} calculations for all three rivet types).

Estimating σ

Because the s chart is in control, the process standard deviation can be estimated for all three rivet types using the formula found in Calculation 14.4.

$$\hat{\sigma} = \frac{\bar{s}}{c_4} = \frac{1.342}{0.9727} = 1.38$$

Calculation 14.4. Estimating σ using $\bar{s}$.

Calculating Process Capability and Performance Ratios

These ratios are calculated using coded data. The coded nominal for the head height characteristic is zero. Therefore, for rivet A, the coded USL is +10 and the coded LSL is −10. Following are calculations for the rivet A head height.

$$C_{pA} = \frac{USL_A - LSL_A}{6\hat{\sigma}} = \frac{10 - (-10)}{6(1.38)} = \frac{20}{8.28} = 2.42$$

Calculation 14.5. C_p calculation for rivet A head height.

$$C_{pk_u A} = \frac{USL_A - \overline{\overline{X}}_A}{3\hat{\sigma}} = \frac{10 - 0.07}{3(1.38)} = \frac{9.93}{4.14} = 2.40$$

Calculation 14.6. $C_{pk\ upper}$ calculation for rivet A head height.

$$C_{pk_l A} = \frac{\overline{\overline{X}}_A - LSL_A}{3\hat{\sigma}} = \frac{0.07 - (-10)}{3(1.38)} = \frac{10.07}{4.14} = 2.43$$

Calculation 14.7. $C_{pk\ lower}$ calculation for rivet A head height.

Target $\overline{X}$ and s Chart Advantages

- Multiple parts or characteristics can be plotted on the same chart (provided they all exhibit similar variability).
- Data from gages that are zeroed out on their target values can be plotted directly on the target $\overline{X}$ without data coding or data transformation.
- Statistical control can be assessed for both the process and each unique part and/or characteristic being made in the process.
- Due to the large subgroup size, the $\overline{X}$ chart is very sensitive to small process shifts.

Target $\overline{X}$ and s Chart Disadvantages

- Requires software to efficiently handle the large amounts of data.
- The use of coded negative numbers can sometimes be confusing.
- When interpreting the target $\overline{X}$ chart, both the zero line and the coded $\overline{\overline{X}}$ must be taken into account. This accounts for some added complexity when interpreting the chart.

Additional Comments About the Case

- Process capability and performance calculations for the B and C rivets are shown in Table 14.7.
- Because the target $\bar{X}$ and s chart proved to be in control, the only values that change when calculating the capability ratios are the specification limits. The coded $\bar{\bar{X}}$ and $\hat{\sigma}$ values used to calculate C_p and C_{pk} ratios are the same for all three rivet types.

Table 14.7. C_p and C_{pk} calculations for B and C rivets.

B rivet	C rivet
$C_{pB} = 1.21$	$C_{pC} = 3.62$
$C_{pk_uB} = 1.19$	$C_{pk_uC} = 3.61$
$C_{pk_lB} = 1.22$	$C_{pk_lC} = 3.64$

Chapter 15

Control Charts for Small Lot Production Runs—The Short Run Charts

In this section, three more control charts will be covered. These special control charts will help the SPC practitioner manage situations where he or she encounters

- Multiple characteristics
- Dissimilar characteristics and standard deviations
- Short production runs
- Limited quantities of data

The control charts found in this chapter are designed to be useful with small lot production runs with limited amounts of data. Like the target charts, data from different production lots are mathematically coded, however, the short run charts go one step further by converting the plot points into unitless ratios. This allows for the use of common control limits that can be used for multiple part numbers or different characteristics. When someone thinks he or she just doesn't have enough data to calculate meaningful control limits, he or she should investigate the short run charts before giving up.

All short run charts share common attributes. Rather than repeating these commonalties in each short run chart chapter, they will be covered once in this overview.

Short run control charts are based on traditional control charts. The three traditional—or core—variables control charts are

- Individual X and moving range chart
- $\overline{X}$ and range chart
- $\overline{X}$ and s chart

All three chart types are used to monitor the stability of a process' central tendency (average) and the variation (standard deviation) about that average. All short run charts in this section are derived from one of these three traditional charts.

Short Run Chart Examples

Short run charts are designed to monitor characteristics with different feature sizes, units of measure, and different standard deviations all on the same control chart. Like target charts, short run charts require data to be mathematically coded. While only averages are coded on a target $\overline{X}$ and range chart™, the short run $\overline{X}$ and range chart™, for example, requires coding of both subgroup average and range values.

Short run charts may be used to monitor all key characteristics on a part regardless of its feature type or differences in standard deviation or averages.

Example 1. A part's profile, hardness, and surface finish can all be tracked on the same short run chart.

Short run charts may also be used to monitor a process' output regardless of the part number or feature type being produced.

Example 2. A five-axis milling machine will cut a multitude of different shaped parts. A single short run chart may be set up to monitor the milling machine's ability to hog, finish, drill, ream, and bore in the X, Y, and/or Z axis.

A single short run chart can even be used to track a part as it travels through its manufacturing operations.

Example 3. A part's key characteristics at each step of the manufacturing operation can be monitored on the same short run chart. That is, one short run chart could be used to track a part starting with pilot hole drilling, then to hogging, finishing, heat treating, straightening, anodizing, painting, curing, and finally part marking.

Short run charts are necessary when characteristics to be monitored on the same chart have different units of measure and/or have different standard deviations. A short run chart could be used to track taper in thousandths of an inch and Rockwell hardness on the same chart (different units), or to track drilled and reamed holes on the same chart (different standard deviations).

As compared to using traditional control charts, using short run control charts will decrease the number of charts that must be managed, while allowing an increase in the number of characteristics that can be tracked.

Short Run Plot Points

Short run plot points are based on the traditional $\overline{X}$, R, and s plot points. In order to monitor dissimilar characteristics on the same chart, the plot points must be coded. This coding of data is what allows different units of measure and different product characteristics to be plotted on the same chart.

Plot Points for the Short Run Range Chart

Before coding, it is helpful to recall how control limits are calculated on the traditional range chart.

Let's assume that for any individual range plot point falling between the control limits found in Equation 15.1, it is said to be in control.

$$\text{UCL}_R = D_4\overline{R} \qquad \text{LCL}_R = D_3\overline{R}$$

Equation 15.1. Upper and lower control limit formulas for traditional *R* chart.

A plot point is in control on an R chart when it falls between the control limits shown in Equation 15.2.

$$\text{UCL}_R > R > \text{LCL}_R$$

or

$$D_4\overline{R} > R > D_3\overline{R}$$

Where R is the actual subgroup range value

Equation 15.2. Traditional *R* chart inequality.

To make the R chart plot points unitless ratios, the $\overline{R}$ must be eliminated from the inequality found in Equation 15.2. To eliminate $\overline{R}$ from the calculations without changing the inequality, simply divide all three terms in Equation 15.2 by $\overline{R}$ (see Equation 15.3).

$$\frac{D_4\overline{R}}{\overline{R}} > \frac{R}{\overline{\overline{R}}} > \frac{D_3\overline{R}}{\overline{R}}$$

Equation 15.3. Dividing traditional R chart inequality by $\overline{R}$.

Canceling the $\overline{R}s$, the result is found in Equation 15.4, in which the new plot point is defined as $\frac{R}{\overline{\overline{R}}}$.

$$D_4 > \frac{R}{\overline{\overline{R}}} > D_3$$

Equation 15.4. Coded plot point and control limits for short run R chart.

The $\overline{R}$ for a given process is the expected or "hoped for" average range, so we will rename it as target $\overline{R}$. Therefore, the short run range plot point is defined in Equation 15.5.

$$\text{Short run range plot point} = \frac{R}{\text{Target }\overline{\overline{R}}}$$

Equation 15.5. Short run range plot point formula.

The short run range plot point is the ratio between an actual subgroup range and an expected or target range. This plot point is a unitless ratio. Neither the chart nor its x-axis is constrained by limitations imposed by plotting data representing different units of measure.

> **Note:** We recognize that, even though the mathematics support plotting different units of measure on the same chart, the rationale for doing so should be founded in legitimacy. Arguably, one may never have a need to mix units of measure on the same chart. However, one may need to mix dissimilar standard deviations on the same chart, such as combining on the same chart similar features from parts made of dissimilar materials.

What Is Target $\overline{R}$?

Target $\overline{R}$ is the heart of the short run data transformation. It represents an estimated or expected range.

How to Estimate Target $\overline{R}$

There are five methods for estimating target $\overline{R}$. Each is described now and listed from most desirable to least desirable.

1. Use $\overline{R}$ from existing in-control range charts.

When using a traditional range control chart, the standard practice is to calculate the centerline and control limits after about 20 plot points. If the range chart is in control, the limits and centerline are extended into the future and used as baselines. New data are plotted against the established baselines. The limits are recalculated only when there has been a sustained change in the process. Therefore, the centerline on an existing in-control range chart can be used as the target $\overline{R}$.

2. Convert existing inspection sampling data into target $\overline{R}$.

If quality assurance personnel have recorded measurements from the characteristics that represent normal production output, the standard deviation of that data can be converted into a target $\overline{R}$ using the formula found in Equation 15.6.

$$\text{Target } \overline{R} = \left(\frac{d_2}{c_4}\right)s$$

where

c_4 is based upon the number of historical measurements

d_2 is based upon the anticipated subgroup size of the short run range control chart

s is the sample standard deviation of the historical data set

Equation 15.6. Formula for calculating target $\overline{R}$ from a historical data set. The derivation for Equation 15.6 can be found in the appendix (Equation A.1).

3. Use $\overline{R}$ from similar characteristics, parts, or process parameters.

If no control charts or quality records exist for a new characteristic to be controlled, but data from a *similar* characteristic exists, use methods 1 or 2 on the similar data to estimate an initial target $\overline{R}$ for the new characteristic.

4. Ask someone who knows the capability of the process. For example, ask the machinist what tolerance the lathe will hold. Suppose the response is, "It will hold ±0.001." Given this statement, one might assume that the machinist was describing the natural variability (the six sigma spread) of the machine's variability. Assuming this, the standard deviation can be estimated by dividing the total tolerance by six. In this example, the estimated standard deviation would be 0.002"/6 = 0.00033". Then use method 2 to convert the estimated s into a target $\overline{R}$.

5. Use engineering tolerance to establish initial target $\overline{R}$. If there is no knowledge of the expected standard deviation of the characteristic, base an initial target $\overline{R}$ on the engineering tolerance using one of the following formulas.

a. For two-sided specifications, use Equation 15.7.

$$\text{Target } \overline{R} = \frac{d_2}{6}(\text{USL} - \text{LSL})$$

Equation 15.7. Formula for calculating target $\overline{R}$ for two-sided specifications. The derivation for Equation 15.7 can be found in the appendix (Equation A.2).

b. For unilateral, or one-sided specifications, use Equation 15.8.

$$\text{Target } \overline{R} = \frac{d_2}{3}\left|\text{Specification limit} - \text{Target } \overline{\overline{X}}\right|$$

Where target $\overline{\overline{X}}$ is the targeted centering of the process

Equation 15.8. Formula for calculating target $\overline{R}$ for unilateral specifications. The derivation for Equation 15.8 can be found in the appendix (Equation A.3).

Note: Method 5 should be used with caution and with full knowledge that a target $\overline{R}$ based on engineering tolerance should not be used as a standard of statistical control. It can, however, be used as a temporary starting point. Once actual data become available, the target $\overline{R}$ should be updated to reflect the new information.

Short Run Range Control Chart

Because the plot points on a short run range control chart are unitless ratios, the UCL is simply D_4, the LCL is D_3, and the centerline is 1 (see Figure 15.1). Therefore, a short run range chart with a sam-

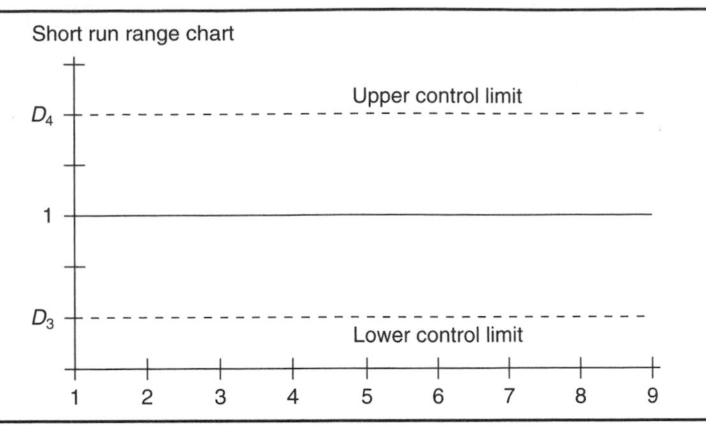

Figure 15.1. D_4 and D_3 are the upper and lower control limits on all short run range charts.

ple size of five would have a UCL = 2.114 and an LCL = 0 (see Table A.1 in the Appendix) with a centerline = 1.

Example

Suppose three parts are manufactured on the same machine. All three parts have different engineering target (nominal) values and different expected standard deviations. Table 15.1 shows the target values based on previous control charts.

The data in Table 15.2 represent measurements from the three parts in subgroups of size five.

The short run range plot points can be generated by taking the range of each subgroup and dividing it by the target $\overline{R}$ for each respective part (see Table 15.3).

Plotting the coded range plot points results in a chart that looks like Figure 15.2.

By coding the range values, parts with different standard deviations can now be plotted on the same control chart. Plotting unitless ratios

Table 15.1. Target $\overline{\overline{X}}$ and target $\overline{R}$ for three parts.

	A	B	C
Target $\overline{\overline{X}}$	17	26	5
Target $\overline{R}$	2.8	6.6	1.6

Table 15.2. Data to be used for completing a short run range chart.

Subgroup number	1	2	3	4	5	6	7	8	9
Part	A	A	A	A	B	B	C	C	C
Sample 1	15.2	15.7	16.6	17.8	26.9	23.1	6.8	4.6	4.6
2	15.5	18.1	17.2	19.3	23.9	25.2	6.9	4.3	4.8
3	18.1	17.3	18.5	20.0	24.0	28.8	4.6	4.4	5.3
4	17.6	15.5	17.7	15.8	22.5	28.2	5.1	4.1	5.4
5	17.6	16.1	17.7	15.9	24.5	25.6	5.5	5.0	5.3
Average	16.8	16.5	17.5	17.8	24.4	26.2	5.8	4.5	5.1
Range	2.9	2.6	1.9	4.2	4.4	5.7	2.3	0.9	0.8

Table 15.3. Calculated plot points for a short run range chart.

Subgroup number	1	2	3	4	5	6	7	8	9
Part	A	A	A	A	B	B	C	C	C
Target $\overline{R}$	2.8	2.8	2.8	2.8	6.6	6.6	1.6	1.6	1.6
R	2.9	2.6	1.9	4.2	4.4	5.7	2.3	0.9	0.8
R/Target $\overline{R}$	1.0	0.9	0.7	1.5	0.7	0.9	1.4	0.6	0.5

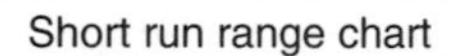

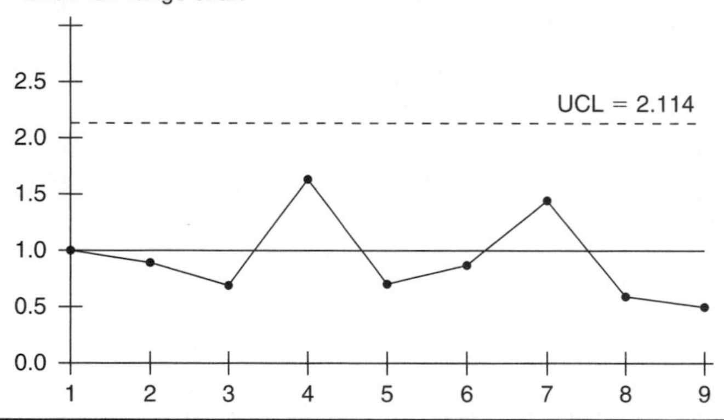

Figure 15.2. Typical short run range chart. UCL = D_4 based on a sample size of five.

may seem a little strange at first, but practitioners soon realize the advantage of evaluating process data in this format.

Plot Points for the Short Run *s* Chart

Traditional *s* chart control limit formulas are found in Equation 15.9.

$$UCL_s = B_4\bar{s} \qquad LCL_s = B_3\bar{s}$$

Equation 15.9. Upper and lower control limit formulas for traditional *s* chart.

For any one plot point falling between the control limits, that plot point is said to be in control. See Equation 15.10 for a definition of in-control plot points for the *s* chart.

$$UCL_s > s > LCL_s$$

or

$$B_4\bar{s} > s > B_3\bar{s}$$

Where s is the sample standard deviation for a given subgroup

Equation 15.10. Traditional *s* chart inequality.

To eliminate $\bar{s}$ from the calculations without changing the inequality, simply divide all three terms by $\bar{s}$ (see Equation 15.11).

$$\frac{B_4\bar{s}}{\bar{s}} > \frac{s}{\bar{s}} > \frac{B_3\bar{s}}{\bar{s}}$$

Equation 15.11. Dividing traditional *s* chart inequality by $\bar{s}$.

Canceling $\bar{s}$ from the inequality results in Equation 15.12.

$$B_4 > \frac{s}{\bar{s}} > B_3$$

Equation 15.12. Coded plot point and control limits for short run *s* chart.

In Equation 15.12, the new plot point is defined as $\frac{s}{\bar{s}}$. If the $\bar{s}$ for a certain process is unknown, then it may be replaced with an expected or target $\bar{s}$.

Therefore, the short run s plot point is expressed in Equation 15.13.

$$\text{Short run } s \text{ plot point} = \frac{s}{\text{Target } \bar{s}}$$

Equation 15.13. Short run s chart plot point formula.

The short run s plot point is the ratio between an actual sample standard deviation and an expected or target s value. This plot point is a unitless ratio. Again, this is an advantage because unitless ratios can be graphed with other unitless ratios on the same chart. Neither the chart nor its x-axis is constrained by limitations imposed by plotting data because of its unit of measure.

Note: We recognize that, even though the mathematics support plotting different units of measure on the same chart, the rationale for doing so should be founded in legitimacy. Arguably, one may never need to mix units of measure on the same chart. However, one may need to mix dissimilar standard deviations on the same chart, such as combining on the same chart similar features from parts made of dissimilar materials.

What Is Target $\bar{s}$?

Target $\bar{s}$ represents an estimated or expected standard deviation and is used to calculate plot points on the short run s chart.

How to Estimate Target $\bar{s}$

There are five methods for estimating target $\bar{s}$. Each is described now and listed from most desirable to least desirable.

1. **Use $\bar{s}$ from existing in-control s charts.**
When using a traditional s control chart, the standard practice is to calculate the centerline and control limits after about 20 plot points. If the chart is in control, the limits and centerline are extended into the future and used as baselines. New data are plotted against the established baselines. The limits are recalculated only when there has been a sustained change in the process. Therefore, the centerline on an existing in-control s chart can be used as the target $\bar{s}$.

2. **Convert existing quality records data into a target $\bar{s}$.**
If quality assurance inspection data from the characteristic exists, and the data represent normal production output, the standard deviation of that data can be converted into a target $\bar{s}$.

A sample standard deviation value is converted into a target $\bar{s}$ value using the formula in Equation 15.14.

$$\text{Target } \bar{s} = \frac{s}{c_4}$$

Where

s is the sample standard deviation based on historical data

c_4 is based on the anticipated sample size of the short run s control chart

Equation 15.14. Converting a sample standard deviation value into a target $\bar{s}$ value. The derivation for Equation 15.14 can be found in the appendix (Equation A.4).

3. Use $\bar{s}$ from similar characteristics, parts, or process parameters.

If no control charts or quality records exist for a new characteristic to be controlled, but data from a *similar* characteristic exists, use methods 1 or 2 on the similar data to estimate an initial target $\bar{s}$ for the new characteristic.

4. Ask someone who knows the capability of the process.

For example, ask the machinist what tolerance the lathe will hold. Suppose the response is, "It will hold ±0.020." The standard deviation can be estimated by dividing the total tolerance by six. In this example, the estimated sample standard deviation would be 0.040"/6 = 0.0067". Assuming that n = 12, equation 15.14 could be applied. The result is 0.0067/.9776 = 0.0069. Therefore, 0.0069" can be used as the target $\bar{s}$.

5. Use engineering tolerance to establish initial target $\bar{s}$.

If there is no knowledge of the expected standard deviation of the characteristic, base an initial target $\bar{s}$ on the engineering tolerance using one of the following formulas.

a. For two-sided specifications, use Equation 15.15.

$$\text{Target } \bar{s} = \frac{c_4}{6}(\text{USL} - \text{LSL})$$

Equation 15.15. Formula for calculating target $\bar{s}$ for two-sided specifications. The derivation for Equation 15.15 can be found in the appendix (Equation A.5).

b. For unilateral or one-sided specifications, use Equation 15.16.

$$\text{Target } \bar{s} = \frac{c_4}{3} \left| \text{Specification limit} - \text{Target } \overline{\overline{X}} \right|$$

Equation 15.16. Formula for calculating target $\bar{s}$ for unilateral specifications. The derivation for Equation 15.16 can be found in the appendix (Equation A.6).

Note: Method 5 should be used with caution and with full knowledge that a target $\bar{s}$ based on engineering tolerance should not be used as a standard of statistical control. It can, however, be used as a temporary starting point. Once actual data become available, the target $\bar{s}$ should be updated to reflect the new information.

Short Run *s* Control Chart

The formula to be used for calculating short run *s* plot points can be found in Equation 15.17.

$$\frac{s}{\text{Target } \bar{s}}$$

Equation 15.17. The plot point equation for a short run *s* chart.

For the short run *s* chart, the UCL is simply B_4, the LCL is B_3, and the centerline is 1 (see Figure 15.3). Therefore, a short run *s* chart with a sample size of 10 would have a UCL = 1.716 and an LCL = 0.284 with a centerline of 1.

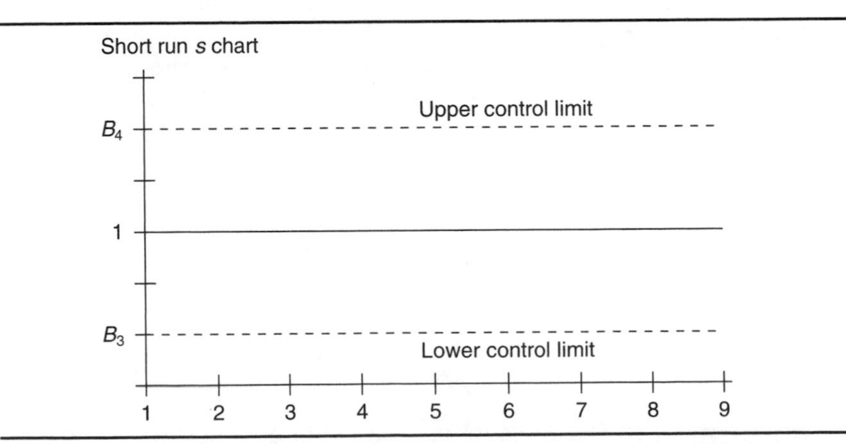

Figure 15.3. B_4 and B_3 are the upper and lower control limits on short run *s* charts.

Example

Suppose there are three parts that are manufactured on the same machine. All three parts have different engineering nominal values and different expected standard deviations. Table 15.4 shows the target values based on previous control charts.

The data in Table 15.5 represent measurements from the three parts in subgroups of size 10.

The plot points for the short run *s* chart can be generated by taking the standard deviation of each subgroup and dividing it by the respective target $\bar{s}$ for each part (see Table 15.6).

Plotting the coded *s* plot points results in Figure 15.4.

By coding the *s* plot points, parts with different standard deviations can now be plotted on the same control chart.

Table 15.4. Target $\bar{\bar{X}}$ and target $\bar{s}$ for three parts.

	A	B	C
Target $\bar{\bar{X}}$	38	14	66
Target $\bar{s}$	1.9	3.2	10.7

Table 15.5. Data to be used for completing a short run *s* chart.

Subgroup number	1	2	3	4	5	6	7	8	9
Part	A	A	A	A	B	B	C	C	C
Sample 1	39.2	38.7	36.8	36.8	12.8	15.4	63.5	57.9	49.2
2	41.1	38.3	38.7	38.5	12.6	13.3	58.5	66.5	69.3
3	38.3	40.6	35.4	36.9	17.5	14.5	72.2	69.8	53.1
4	39.6	39.1	38.0	36.6	14.4	14.3	58.1	68.4	57.1
5	38.2	37.7	36.2	38.2	14.5	15.4	40.9	67.1	63.2
6	34.8	40.0	36.1	38.8	7.1	14.4	62.3	64.2	73.7
7	39.2	41.3	37.6	37.6	15.2	17.9	69.9	90.5	59.2
8	38.2	38.7	37.9	39.4	9.7	16.0	64.7	75.3	69.5
9	36.8	35.5	37.0	36.9	14.1	13.7	66.0	65.9	55.2
10	43.4	39.5	38.2	41.0	20.6	11.3	78.8	62.5	71.2
Average	38.9	38.9	37.2	38.1	13.9	14.6	63.5	68.8	62.1
s	2.3	1.6	1.1	1.4	3.8	1.8	10.1	8.9	8.5

Table 15.6. Calculated plot points for a short run *s* chart.

Subgroup number	1	2	3	4	5	6	7	8	9
Part	A	A	A	A	B	B	C	C	C
Target $\bar{s}$	1.9	1.9	1.9	1.9	3.2	3.2	10.7	10.7	10.7
s	2.3	1.6	1.1	1.4	3.8	1.8	10.1	8.9	8.5
s/Target $\bar{s}$	1.2	0.8	0.6	0.7	1.2	0.6	0.9	0.8	0.8

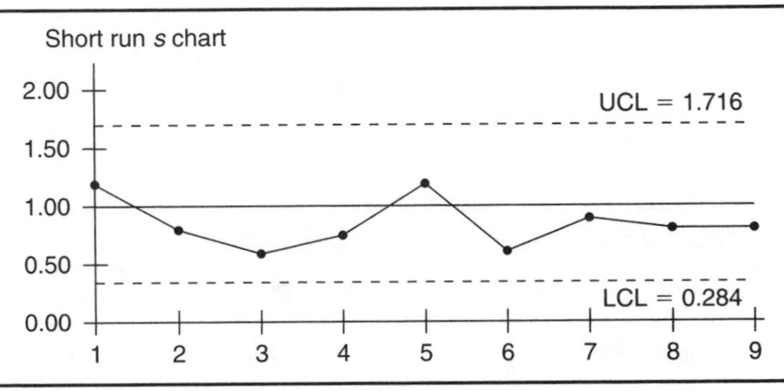

Figure 15.4. Typical short run *s* chart. Upper and lower control limits based on a sample size of 10.

Plot Points for the Short Run $\bar{X}$ Chart

Start with the traditional $\bar{X}$ control limit formulas in Equation 15.18.

$$UCL_{\bar{X}} = \bar{\bar{X}} + A_2\bar{R} \qquad LCL_{\bar{X}} = \bar{\bar{X}} - A_2\bar{R}$$

Equation 15.18. Upper and lower control limits for traditional control charts.

A plot point (an average) is in control when it falls between $\bar{X}$ chart control limits as defined in Equation 15.19.

$$UCL_{\bar{X}} > \bar{X} > LCL_{\bar{X}}$$

or

$$\bar{\bar{X}} + A_2\bar{R} > \bar{X} > \bar{\bar{X}} - A_2\bar{R}$$

Equation 15.19. Traditional $\bar{X}$ chart inequality.

With the short run $\bar{X}$ chart, both $\bar{\bar{X}}$ and $\bar{R}$ need to be removed from the inequality so that only A_2 remains. Doing this will result in a short run $\bar{X}$ chart whose control limits are $-A_2$ and $+A_2$. To do this without changing the inequality, one must first subtract $\overline{\overline{X}}$ from all three terms (see Equation 15.20).

$$(\overline{\overline{X}} + A_2\overline{R}) - \overline{\overline{X}} > \overline{X} - \overline{\overline{X}} > (\overline{\overline{X}} - A_2\overline{R}) - \overline{\overline{X}}$$

Equation 15.20. Subtraction of $\overline{\overline{X}}$ from traditional $\overline{X}$ chart inequality.

The result of Equation 15.20 is found in Equation 15.21.

$$+A_2\overline{R} > \overline{X} - \overline{\overline{X}} > -A_2\overline{R}$$

Equation 15.21. Result of subtracting $\overline{\overline{X}}$ from traditional $\overline{X}$ chart inequality.

Next, $\overline{R}$ must be eliminated from the inequality. This is done by dividing the inequality by $\overline{R}$ (see Equation 15.22).

$$\frac{+A_2\overline{R}}{\overline{R}} > \frac{\overline{X} - \overline{\overline{X}}}{\overline{R}} > \frac{-A_2\overline{R}}{\overline{R}}$$

Equation 15.22. Division of traditional $\overline{X}$ chart inequality by $\overline{R}$.

Canceling the $\overline{R}$*s* produces the result found in Equation 15.23.

$$+A_2 > \frac{\overline{X} - \overline{\overline{X}}}{\overline{R}} > -A_2$$

Equation 15.23. Coded plot point and control limits for short run $\overline{X}$ control chart.

The short run $\overline{X}$ plot point is shown in Equation 15.24.

$$\text{Short run } \overline{X} \text{ plot point} = \frac{\overline{X} - \text{Target } \overline{\overline{X}}}{\text{Target } \overline{R}}$$

Equation 15.24. Formula for the short run $\overline{X}$ plot point (to be used in conjunction with a short run R chart).

The short run $\overline{X}$ plot point is the ratio between a coded $\overline{X}$ (deviation from target $\overline{\overline{X}}$) and a target range (target $\overline{R}$).

Note: Short run *IX* plot points using *MR* and short run $\overline{X}$ plot points using *s* are calculated in a similar fashion as short run $\overline{X}$ plot points. See Equations 15.25 and 15.26.

$$\text{Short run individual } X \text{ plot point} = \frac{IX - \text{Target } \overline{IX}}{\text{Target } \overline{MR}}$$

Equation 15.25. Formula for the short run *IX* chart plot point (to be used in conjunction with a short run *MR* chart).

$$\text{Short run } \overline{X} \text{ plot point} = \frac{\overline{X} - \text{Target } \overline{\overline{X}}}{\text{Target } \bar{s}}$$

Equation 15.26. Formula for the short run $\overline{X}$ plot point (to be used in conjunction with a short run *s* chart).

What Is Target $\overline{\overline{X}}$?

Target $\overline{\overline{X}}$ is the targeted or expected average of a process parameter or characteristic.

How to Estimate Target $\overline{\overline{X}}$

There are four ways to estimate target $\overline{\overline{X}}$, each numbered from most desirable to least desirable.

1. **Use $\overline{\overline{X}}$ from existing in-control $\overline{X}$ charts.**
   When using a traditional $\overline{X}$ chart, the standard practice is to calculate the $\overline{\overline{X}}$ and control limits after about 20 plot points. If the control chart is in control, the limits and centerline are used as baselines. They are extended into the future and current data are plotted against the established centerline and control limits. The limits are recalculated only when there has been a sustained change in the process. Therefore, the centerline on an existing in-control $\overline{X}$ chart can be used as the target $\overline{\overline{X}}$.
2. **Convert existing quality assurance sampling data into target $\overline{\overline{X}}$.**
   If quality assurance inspection data from the characteristic exist, and the data represent normal production output, the average of that data can be used as the target $\overline{\overline{X}}$.
3. **Use $\overline{\overline{X}}$ from similar characteristics, parts, or process parameters.**
   If no control charts or quality records exist for the new characteristic to be controlled, but there exists charts or quality records from a *similar* characteristic, use methods 1 or 2 on the similar data to estimate an initial target $\overline{\overline{X}}$ for the new characteristic.

4. Use engineering print nominal as target $\bar{\bar{X}}$.

If there is no knowledge of the expected centering of the characteristic to be monitored, base the initial target $\bar{\bar{X}}$ on engineering nominal (the midpoint between the USL and LSL).

For unilateral tolerances, pick a target value sufficiently (preferably greater than three standard deviations) away from the specification to ensure minimal fallout.

Short Run $\bar{X}$ Control Chart

The plot points on a short run $\bar{X}$ control chart are unitless ratios. The formula for calculating the plot points can be found in Equation 15.24.

Recall that, for any short run $\bar{X}$ control chart, the UCL will be equal to $+A_2$, the LCL will be equal to $-A_2$ and the centerline will always have a value of zero (see Figure 15.5).

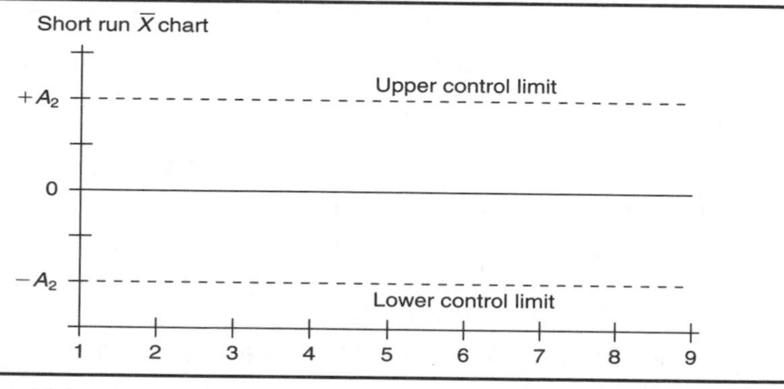

Figure 15.5. $\underline{+}A_2$ and $-A_2$ are the upper and lower control limits on all short run $\bar{X}$ charts used in conjunction with short run $\bar{R}$ charts.

Example

Suppose there are three parts that are manufactured on the same machine. All three parts have different engineering nominal values and different expected standard deviations. Table 15.7 shows the target values based on previous control charts.

Table 15.7. Target $\bar{\bar{X}}$ and target $\bar{R}$ for three parts.

	A	B	C
Target $\bar{\bar{X}}$	17.0	26.0	5.0
Target $\bar{R}$	2.8	6.6	1.6

Table 15.8. Average and range values to be used in calculating short run $\overline{X}$ chart plot points.

Subgroup number	1	2	3	4	5	6	7	8	9
Part	A	A	A	A	B	B	C	C	C
Sample 1	15.2	15.7	16.6	17.8	26.9	23.1	6.8	4.6	4.6
2	15.5	18.1	17.2	19.3	23.9	25.2	6.9	4.3	4.8
3	18.1	17.3	18.5	20.0	24.0	28.8	4.6	4.4	5.3
4	17.6	15.5	17.7	15.8	22.5	28.2	5.1	4.1	5.4
5	17.6	16.1	17.7	15.9	24.5	25.6	5.5	5.0	5.3
Average	16.8	16.5	17.5	17.8	24.4	26.2	5.8	4.5	5.1
Range	2.9	2.6	1.9	4.2	4.4	5.7	2.3	0.9	0.8

Table 15.9. Calculated plot points (bottom row) for a short run $\overline{X}$ chart.

Subgroup number	1	2	3	4	5	6	7	8	9
Part	A	A	A	A	B	B	C	C	C
$\overline{X}$	16.8	16.5	17.5	17.8	24.4	26.2	5.8	4.5	5.1
Target $\overline{\overline{X}}$	17.0	17.0	17.0	17.0	26.0	26.0	5.0	5.0	5.0
$\overline{X}$ − Target $\overline{\overline{X}}$	−0.2	−0.5	0.5	0.8	−1.6	0.2	0.8	−0.5	0.1
Target $\overline{R}$	2.8	2.8	2.8	2.8	6.6	6.6	1.6	1.6	1.6
$\dfrac{\overline{X} - \text{Target } \overline{\overline{X}}}{\text{Target } \overline{R}}$	−0.07	−0.18	0.18	0.29	−0.24	0.03	0.5	−0.31	0.06

The nine subgroups in Table 15.8 represent measurements from the three parts in subgroups of size five.

The short run $\overline{X}$ chart plot points can be generated by taking the average and range of each subgroup in Table 15.8 and using them in the formula for the short run $\overline{X}$ plot point (see Equation 15.24). The result can be found in Table 15.9.

The short run $\overline{X}$ points can then be put into a graphical control chart format like Figure 15.6.

Benefits of Using Unitless Ratios as Plot Points

Although plotting unitless ratios may seem a little strange at first, several very important advantages can be realized, including

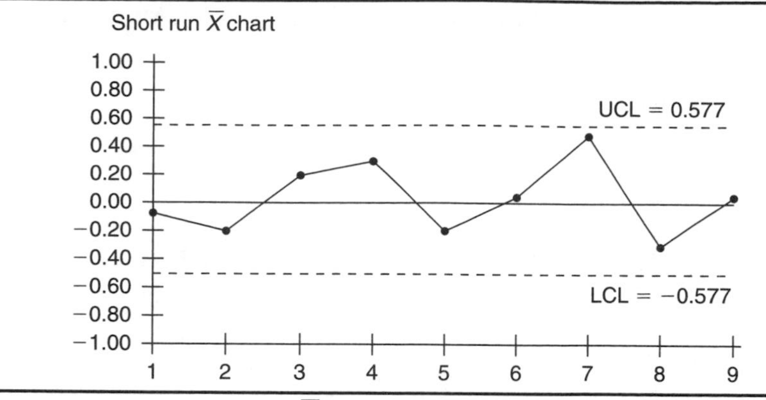

Figure 15.6. A short run $\overline{X}$ chart using plot points found in Table 15.9.

1. Control limits and centerlines never need to be calculated or recalculated. They are constants based only on sample size.
2. Control limits are known before the first point is plotted.
3. Because the scale is not related to the actual units of measurement, the interpreter can focus solely on the positioning of plot points relative to each other, the control limits, and the centerline.

Interpreting Short Run Charts

Short run charts are interpreted in a fashion similar to their traditional chart counterparts. The only feature that makes short run chart interpretation different is having to know from where the target values come. If the targets are based on previous in-control charts, there is little difference between short run and traditional chart interpretation.

When target values are estimated from historical data, similar parts, or engineering tolerances, the charts are initially analyzed to determine if the estimated target values are accurate. If the charts initially appear out of control, it may be interpreted as an out-of-control situation or that the target values are not accurate. In either case, there should be an investigation for the cause of the control chart signal. Either the process has changed significantly or the target values simply were initially inaccurate and require updating. If the charts initially appear to be in control, the target values are assumed to be accurate and no updating is required.

Out-of-Control Conditions

Once the target values are determined to be accurate, out-of-control conditions for the process should be investigated. Out-of-control conditions are detected using the following standard criteria.

- Plot points beyond control limits
- Runs above or below the centerline
- Any obvious pattern (trends, shifts, or cycles)

Detecting Process Improvements on the Short Run Range and *s* Charts

A run below the centerline on a range or *s* chart is an indication of an improvement (reduced variation). When this occurs, only the target $\bar{R}$s and/or target $\bar{s}$s need to be updated—recalculation of control limits is not necessary because they are constants. Targets are updated by averaging the actual subgroup ranges or standard deviations that are representative of the process improvement. Target values should be updated only if the reason for the improvement detected on the chart can be identified and there are sufficient data to confirm a sustained improvement.

Detecting Process Improvement on the Short Run $\bar{X}$ Chart

Typically, the target $\bar{\bar{X}}$ is the engineering nominal or some value of desired centering. Improvements, therefore, are detected when the plot points exhibit less variability about their target values (the zero line).

When to Update Target Values

Target values should be updated when

1. Initial target estimates are found to be inaccurate.
2. A process improvement occurs and the cause is identified.

Note: Target values should not be updated to reflect a deterioration in the process unless it is an intentional and/or permanent condition.

Chapter 16

Short Run Individual *X* and Moving Range Chart

Decision Tree Section

Table 16.1. Control chart decision tree.

Number of characteristics or locations on the same chart?	Subgroup size?	Different averages on the same chart?	Different standard deviations on the same chart?	Use this chart
1	1	Yes	Yes	Short run *IX* and moving range

Description

Short Run Individual X *Chart*

The short run individual X chart is used to monitor and detect changes in individual measurements among characteristics of any type. The characteristics may have different nominals, different units of measure, and different standard deviations, but should be related enough to want to analyze them all on the same chart. The plot points represent individual measurements that are coded by subtracting a target $\overline{IX}$ (usually an engineering nominal value) from each measurement and then dividing the result by a target $\overline{MR}$. Each characteristic on the chart may have a unique target $\overline{IX}$ and target $\overline{MR}$.

These charts are mostly used to monitor characteristics where only one measurement is necessary to represent a process at a given period of time. Examples include accounting values, homogeneous batches such as concentration in a chemical bath, and characteristics that, due to the nature of the process, the sources of variation change significantly from one sampling opportunity to the next.

Short Run Moving Range Chart

The short run moving range chart is used to monitor and detect changes in the standard deviations among characteristics of any type. The plot points on the short run moving range chart represent the absolute difference between consecutive coded short run *IX* chart plot points.

Subgroup Assumptions

- Independent measurements
- Constant sample size

Calculating Plot Points

Table 16.2. Formulas for calculating plot points for the short run *IX* and *MR* chart.

Chart	Plot point	Plot point formula
Short run individual *X*	Coded *IX*	$\dfrac{IX - \text{Target } \overline{IX}}{\text{Target } \overline{MR}}$
Short run moving range	Coded MR	Absolute difference between two consecutive coded *IX* values from the same part or location

Calculating Centerlines

Table 16.3. Centerlines on the short run *IX* and *MR* chart.

Chart	Centerline	Centerline formula
Short run individual *X*	0	0
Short run moving range	1	1

Calculating Control Limits

Table 16.4. Control limits on the short run *IX* and *MR* chart.

Chart	Upper control limit	Lower control limit
Short run individual *X*	$+A_2$	$-A_2$
Short run moving range	D_4	D_3

Example

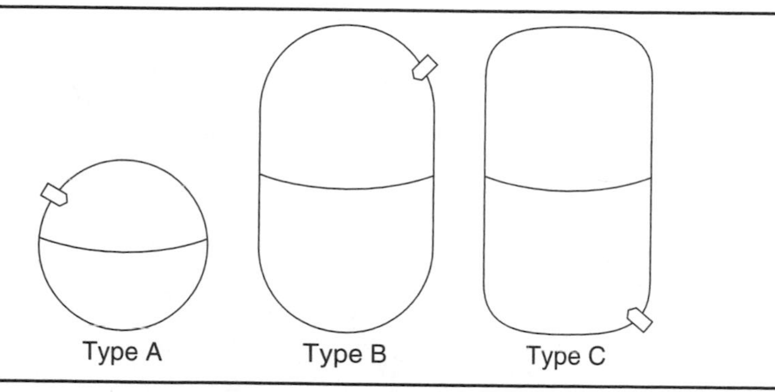

Figure 16.1. Three fire extinguishing bottles, each with different burst pressure requirements.

Case Description

A certain manufacturer of aerospace fire extinguishing bottles performs destructive testing on each batch of bottles. The test involves pressurizing the bottle until it bursts. Burst pressure is the key characteristic. Each bottle's burst requirements are different. Also, since each bottle type can be made of different materials with different wall thickness, burst pressure variability changes with each bottle type. For these reasons, a short run *IX–MR* chart is selected to monitor all data from the burst test. All target values were obtained from past control charts.

Table 16.5. Target values and minimum specification limit for all three bottle types.

Bottle type	Target $\overline{IX}$	Target $\overline{MR}$	Minimum specification limit
Type A	1162	22.6	1070
Type B	678	13.6	625
Type C	603	10.2	550

Sampling Strategy

Since burst testing is destructive, only one bottle from each lot is tested—typically the first piece. However, results from all burst tests are recorded on the same control chart. Tests are immediately performed as first-piece bottles become available. One test stand supports the entire manufacturing operation. Bottle types can change for each test.

Data Collection Sheet

Table 16.6. Burst test data including plot point calculations for the short run *IX* and *MR* chart.

Subgroup number	1	2	3	4	5	6	7	8	9
Bottle type	A	C	C	A	A	B	C	B	A
Burst pressure	1156	576	600	1136	1169	666	609	692	1188
MR	—	—	24	20	33	—	9	26	19
Target $\overline{IX}$	1162	603	603	1162	1162	678	603	678	1162
$IX -$ Target $\overline{IX}$	−6	−27	−3	−26	7	−12	6	14	26
Target $\overline{MR}$	22.6	10.2	10.2	22.6	22.6	13.6	10.2	13.6	22.6
$\frac{IX - \text{Target } \overline{IX}}{\text{Target } \overline{MR}}$	−0.27	−2.65	−0.29	−1.15	0.31	−0.88	0.59	1.03	1.15
Coded MR	—	—	2.36	0.88	1.46	—	0.88	1.91	0.84

Subgroup number	10	11	12	13	14	15	16	17	18
Bottle type	B	B	C	A	C	C	A	B	B
Burst pressure	674	686	604	1186	615	600	1197	690	699
MR	18	12	5	2	11	15	11	4	9
Target $\overline{IX}$	678	678	603	1162	603	603	1162	678	678
$IX -$ Target $\overline{IX}$	−4	8	1	24	12	−3	35	12	21
Target $\overline{MR}$	13.6	13.6	10.2	22.6	10.2	10.2	22.6	13.6	13.6
$\frac{IX - \text{Target } \overline{IX}}{\text{Target } \overline{MR}}$	−0.29	0.59	0.10	1.06	1.18	−0.29	1.55	0.88	1.54
Coded MR	1.32	0.88	0.49	0.09	1.08	1.47	0.49	0.29	0.66

Table 16.6. Cont.

Subgroup number	19	20	21	22	23	24	25	26	27
Bottle type	C	C	B	B	C	B	A	A	A
Burst pressure	602	611	685	679	599	690	1139	1157	1184
MR	2	9	14	6	12	11	58	18	27
Target $\overline{IX}$	603	603	678	678	603	678	1162	1162	1162
IX – Target $\overline{IX}$	−1	8	7	1	−4	12	−23	−5	22
Target $\overline{MR}$	10.2	10.2	13.6	13.6	10.2	13.6	22.6	22.6	22.6
$\frac{IX - \text{Target } \overline{IX}}{\text{Target } \overline{MR}}$	−0.10	0.78	0.51	0.07	−0.39	0.88	−1.02	−0.22	0.97
Coded *MR*	0.19	0.88	1.03	0.44	1.17	0.81	2.57	0.8	1.19

Note: The MR and coded MR values found in Table 16.6 are calculated using previous data points from the same bottle type. For example, the coded *MR* value of 0.49 in subgroup 16 is the result of taking the absolute difference between the coded *IX* values in subgroups 13 and 16: $|1.06 - 1.55| = 0.49$.

Short Run IX *and* MR *Chart*

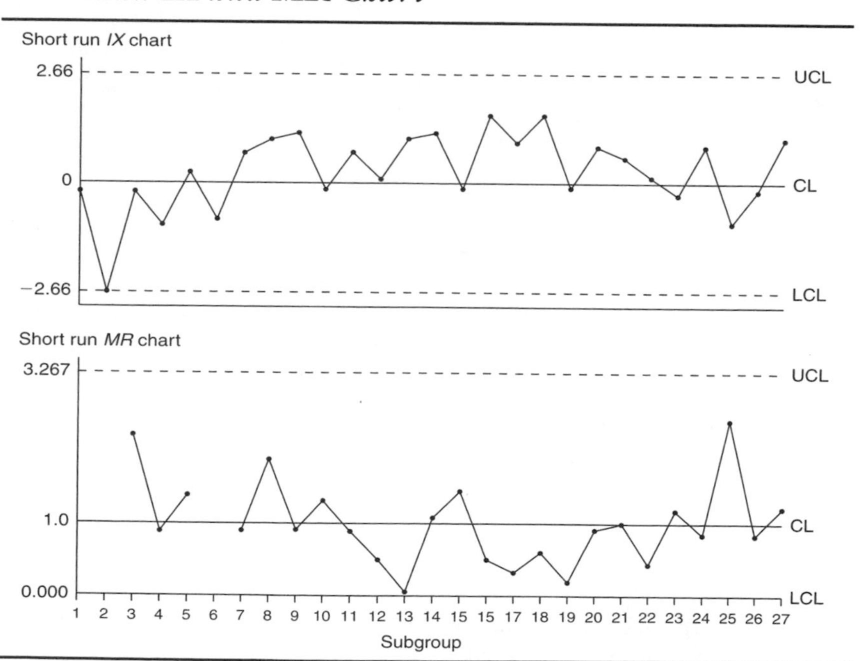

Figure 16.2. Bottle burst test data short run *IX–MR* control chart.

Chart Interpretation

Short run moving range chart: Because there are no nonrandom patterns or points outside control limits, the variability in burst pressure is consistent across all three bottle types.
Short run IX *chart:* The individual plot points appear to be stable with no nonrandom patterns occurring.

Recommendation

Because both charts are in control, the target values (obtained from past control charts) are still appropriate for the current data. Continue maintaining the control chart with no changes in target values.

Estimating the Process Average

Estimates of the process average should be calculated separately for each characteristic or part on the short run *IX* and *MR* chart. In this case, estimates of the process average should be calculated separately for each bottle type. This is illustrated with bottle type A in Calculation 16.1.

$$\overline{IX}_{\text{type A}} = \frac{\sum IX_{\text{type A}}}{k_{\text{type A}}} = \frac{10{,}512}{9} = 1168.0$$

Calculation 16.1. Estimate of average burst pressure for bottle type A.

Estimating σ

Estimates of σ are also calculated separately for each characteristic or location represented on short run *IX* and *MR* charts. In this case, estimates of the standard deviation should be calculated for each bottle type. The calculation of $\overline{MR}$ for bottle type A is cound in Calculation 16.2.

$$\overline{MR}_{\text{type A}} = \frac{\sum MR_{\text{type A}}}{k_{\text{type A}} - 1} = \frac{188}{8} = 23.5$$

Calculation 16.2. Calculation of the average moving range for bottle type A (to be used in estimating its standard deviation).

$$\hat{\sigma}_{\text{type A}} = \frac{\overline{MR}_{\text{type A}}}{d_2} = \frac{23.5}{1.128} = 20.83$$

Calculation 16.3. Estimate of the process standard deviation for bottle type A.

Note: To ensure reliable estimates, k needs to be at least 20. For bottle type A, k is only 9. Therefore, the estimates here and in Table 16.7 are used for illustration purposes only.

Calculating Process Capability and Performance Ratios

Recall that the minimum specification for bottle type A burst pressure is 1070. Because there is only a single minimum burst specification, C_p and $C_{pk\,\text{upper}}$ are not calculated.

$$C_{pk_l\,\text{type A}} = \frac{\overline{IX}_{\text{type A}} - \text{LSL}}{3\hat{\sigma}_{\text{type A}}} = \frac{1168.0 - 1070}{3(20.83)} = \frac{98.0}{62.49} = 1.57$$

Calculation 16.4. $C_{pk\,\text{lower}}$ calculation for bottle type A burst pressure.

Short Run IX–MR *Chart Advantages*

- Graphically illustrates the variation of multiple product or process characteristics on the same chart.
- Can chart process parameters that have changing target values.
- Characteristics from different parts with different means, different standard deviations, and different units of measure can be analyzed on the same chart.
- Pinpoints the characteristics that are in need of the most attention.
- Separates variation due to the process from variation that is product specific.

Short Run IX–MR *Chart Disadvantages*

- The MR chart is dependent upon consecutive IX chart plot points.
- $\overline{IX}$, $\overline{MR}$, and estimates of σ must be calculated separately for each characteristic on the chart.

Additional Comments About the Case

- The case study shown here displayed three bottle types. In the actual situation, there were 22 different bottle types being monitored on the same short run *IX–MR* chart.
- Process capability and performance calculations for the remaining bottle types are shown in Table 16.7.

Table 16.7. Additional summary statistics and process capability and performance ratios for remaining bottle types.

Type B	Type C
$\overline{IX}_{\text{type B}} = 684.6$	$\overline{IX}_{\text{type C}} = 601.8$
$\overline{MR}_{\text{type B}} = 12.5$	$\overline{MR}_{\text{type C}} = 10.9$
$\hat{\sigma}_{\text{type B}} = 11.08$	$\hat{\sigma}_{\text{type C}} = 9.64$
$C_{pk_l\ \text{type B}} = 1.79$	$C_{pk_l\ \text{type C}} = 1.79$

Chapter 17

Short Run $\overline{X}$ and Range™ Chart

Decision Tree Section

Table 17.1. Control chart selection tree.

Number of characteristics or locations on the same chart?	Subgroup size?	Different averages on the same chart?	Different standard deviations on the same chart?	Use this chart
1	>1 <10	Yes	Yes	Short run $\overline{X}$ and range™

Description

Short Run $\overline{X}$ *Chart*

The short run $\overline{X}$ chart is used to monitor and detect changes in the averages among multiple characteristics of any type. The characteristics may have different nominals, different units of measure, and different standard deviations. However, all characteristics on the chart should be related enough to warrant analyzing them together. The plot points are coded by subtracting from each subgroup average its respective target $\overline{\overline{X}}$ (usually the engineering nominal value), and then dividing by its target $\overline{R}$. Each characteristic on the chart has its own unique target $\overline{\overline{X}}$ and target $\overline{R}$. For each characteristic, the number of measurements taken in a subgroup may range from two to nine, but practitioners generally use subgroup sizes of three to five.

Short Run Range Chart

The short run range chart is used to monitor and detect changes in the standard deviations among multiple characteristics. The plot points are coded by dividing the subgroup range by its respective target $\overline{R}$.

Subgroup Assumptions

- Independent measurements
- Constant sample size

Calculating Plot Points

Table17.2. Formulas for calculating plot points for the short run $\overline{X}$ and range™ chart.

Chart	Plot point	Plot point formula
Short run $\overline{X}$	Coded $\overline{X}$	$\dfrac{\overline{X} - \text{Target } \overline{\overline{X}}}{\text{Target } \overline{R}}$
Short run range	Coded R	$\dfrac{R}{\text{Target } \overline{R}}$

Calculating Centerlines

Table 17.3. Centerlines on the short run $\overline{X}$ and range™ chart.

Chart	Centerline	Centerline formula
Short run $\overline{X}$	0	0
Short run range	1	1

Calculating Control Limits

Table 17.4. Control limits on the short run $\overline{X}$ and range™ chart.

Chart	Upper control limit	Lower control limit
Short run $\overline{X}$	$+A_2$	$-A_2$
Short run range	D_4	D_3

Example

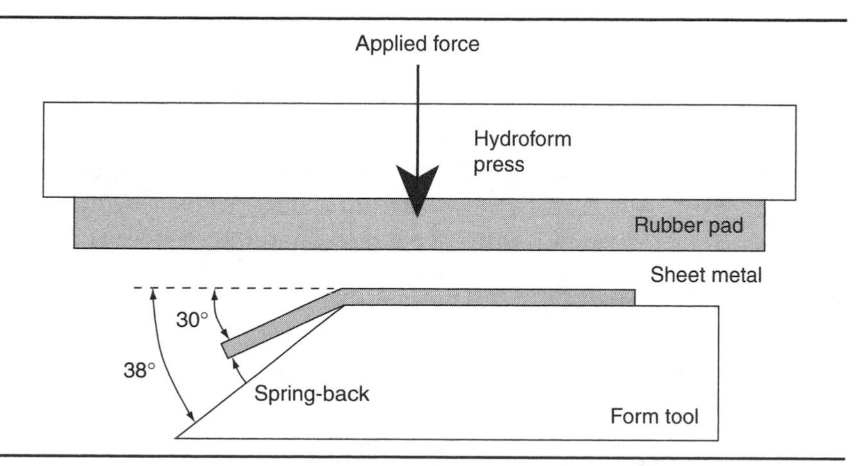

Figure 17.1. Example of sheet metal spring-back after hydroform operation.

Case Description

A hydroform is used to form angles in sheet metal. This is done by compressing a piece of sheet metal between a rubber pad and a form tool. When the metal is bent on the form tool, it springs back a few degrees when the pressure is released. To counteract the spring-back effect, the form tool angle exceeds the desired angle. In this case, the desired resultant sheet metal angles are 30°, 45°, and 90°. The average spring-back and standard deviations are different for each angle. The production foreman wants to use one control chart to monitor the spring-back behavior of all three types of angles. Table 17.5 shows the spring-back target values and specifications.

Table 17.5. Spring-back target values and specifications for three types of angles.

	Angle		
	30°	45°	90°
Spring-back target $\overline{\overline{X}}$	8.2°	4.0°	1.3°
Spring-back target $\overline{R}$	0.35°	0.23°	0.19°
Engineering specifications	8.2° ± 0.5°	4.0° ± 0.5°	1.3° ± 0.5°

Note: The target $\overline{\overline{X}}$ values are based on engineering nominal values and the target $\overline{R}$ values are based on historical quality records.

Sampling Strategy

The hydroform machine is initially set up to bend 45° angles. Five consecutive spring-back measurements are taken every hour until the job is complete. Next, the machine is set up to run 30° angles and so on. Sampling continues in the same manner as before. All measurements are plotted on the same short run $\overline{X}$ and range™ chart.

Data Collection Sheet

Table 17.6. Spring-back data including short run plot point calculations.

Angle	45°	45°	45°	30°	30°	30°	30°	30°
Subgroup number	1	2	3	4	5	6	7	8
1	3.95	3.96	3.91	8.38	8.48	8.33	8.23	8.70
2	3.99	4.08	4.12	8.36	8.52	8.44	8.40	8.76
3	4.14	3.73	4.02	8.49	8.37	8.53	8.26	8.14
4	4.02	4.15	3.76	8.35	8.48	8.34	8.30	7.87
5	4.08	4.07	3.98	8.46	8.16	8.31	8.19	8.27
$\overline{X}$	4.04	4.00	3.96	8.41	8.40	8.39	8.28	8.35
Target $\overline{\overline{X}}$	4.00	4.00	4.00	8.20	8.20	8.20	8.20	8.20
$\overline{X}$ − Target $\overline{\overline{X}}$	0.04	0.00	−0.04	0.21	0.20	0.19	0.08	0.15
Target $\overline{R}$	0.23	0.23	0.23	0.35	0.35	0.35	0.35	0.35
$\frac{\overline{X} - \text{Target } \overline{\overline{X}}}{\text{Target } \overline{R}}$	0.16	−0.01	−0.18	0.59	0.58	0.54	0.22	0.42
R	0.19	0.42	0.36	0.14	0.36	0.22	0.21	0.89
R/Target $\overline{R}$	0.83	1.83	1.57	0.40	1.03	0.63	0.60	2.54

Table 17.6. Cont.

Angle	30°	90°	90°	30°	30°	30°	30°	30°
Subgroup number	9	10	11	12	13	14	15	16
1	8.31	1.19	1.25	8.27	8.07	8.60	8.53	8.38
2	8.30	1.17	1.21	7.87	8.70	8.66	8.74	8.36
3	8.33	1.12	1.21	8.35	8.60	8.65	8.62	8.56
4	8.53	1.17	1.22	8.50	8.28	8.57	8.64	8.29
5	8.32	1.19	1.21	8.74	8.27	8.51	7.98	8.50
$\overline{X}$	8.36	1.17	1.22	8.35	8.38	8.60	8.50	8.42
Target $\overline{\overline{X}}$	8.20	1.30	1.30	8.20	8.20	8.20	8.20	8.20
$\overline{X}$ − Target $\overline{\overline{X}}$	0.16	−0.13	−0.08	0.15	0.18	0.40	0.30	0.22
Target $\overline{R}$	0.35	0.19	0.19	0.35	0.35	0.35	0.35	0.35
$\dfrac{\overline{X} - \text{Target } \overline{\overline{X}}}{\text{Target } \overline{R}}$	0.45	−0.69	−0.42	0.42	0.53	1.14	0.86	0.62
R	0.23	0.07	0.04	0.87	0.63	0.15	0.76	0.27
R/Target $\overline{R}$	0.66	0.37	0.21	2.49	1.80	0.43	2.17	0.77

Angle	45°	45°	45°	90°
Subgroup number	17	18	19	20
1	3.79	3.83	3.96	1.24
2	4.01	4.08	4.08	1.20
3	3.81	3.80	4.04	1.20
4	4.11	3.99	3.95	1.24
5	4.00	3.89	4.04	1.19
$\overline{X}$	3.94	3.92	4.01	1.21
Target $\overline{\overline{X}}$	4.00	4.00	4.00	1.30
$\overline{X}$ − Target $\overline{\overline{X}}$	−0.06	−0.08	0.01	−0.09
Target $\overline{R}$	0.23	0.23	0.23	0.19
$\dfrac{\overline{X} - \text{Target } \overline{\overline{X}}}{\text{Target } \overline{R}}$	−0.24	−0.36	0.06	−0.45
R	0.32	0.28	0.13	0.05
R/Target $\overline{R}$	1.39	1.22	0.57	0.26

Short Run $\overline{X}$ and Range™ Chart

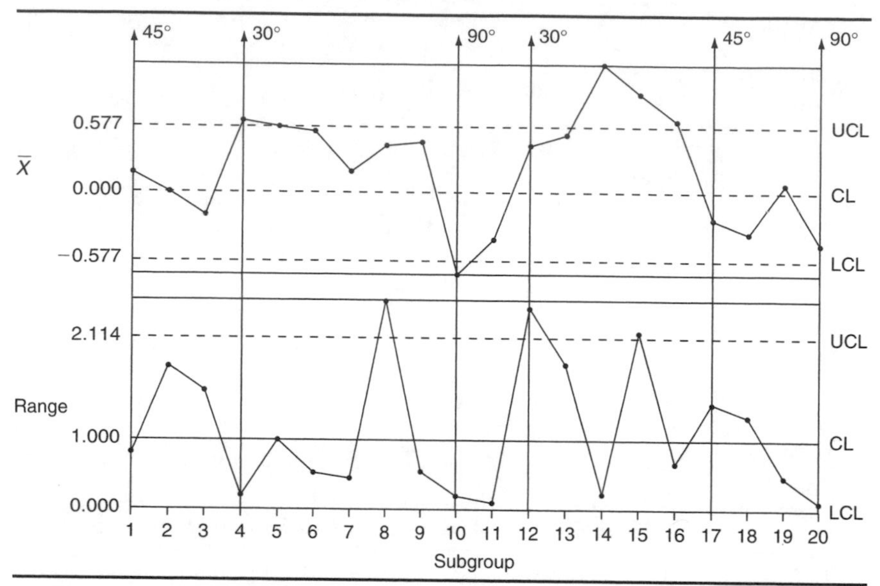

Figure 17.2. Spring-back short run $\overline{X}$ and range™ control charts.

Chart Interpretation

Short run range chart: Three 30° plot points fall above the UCL and are an indication that the variability for the 30° bends is greater than expected. The 45° plot points appear to be behaving randomly. The 90° plot points all fall below the centerline. Each pattern appears to be unique to each bend angle. There appear to be no visible patterns or trends that consistently appear across all bend angles collectively.

Short run $\overline{X}$ chart: All 11 30° plot points fall above the centerline and five fall above the UCL. This indicates that the actual spring-back on 30° bends is greater than the established 8.2° target value. The 45° plot points appear to vary randomly about their target value.

The 90° plot points all fall below the centerline with one of them falling below the LCL. This indicates that the actual spring-back on 90° bends is less than the target $\overline{\overline{X}}$ value of 1.3°. All plot point patterns appear unique to each bend angle. No trends are apparent across all bend angles collectively.

Recommendations

30° Bend Angles

Range plot points erratically jumping above the UCL generally indicate unstable short-term variation. This might be caused by a process change that happens to occur within a subgroup. To pinpoint the cause, a 100 percent sampling strategy with a sample size of one may need to be temporarily established.

The average spring-back is consistently greater than the established target $\overline{\overline{X}}$ of 8.2°. Investigate why the spring-back rates are so much larger than the engineering target and improve the process' ability to maintain a lesser spring-back.

45° Bend Angles

Both ranges and averages appear to behave with consistent variability. The control chart reveals no specific process control issues that need to be addressed with respect to this bend angle.

90° Bend Angles

There are only three plot points on the short run chart that represent the 90° bend angles being produced (subgroups 10, 11, and 20). However, two of the three plot points on the short run $\overline{X}$ chart are very close to the LCL and one falls below. If all three subgroups were consecutive, the two-out-of-three zone analysis rule would be triggered. The user of the control chart should try to find an obvious reason for the low bend angle values. If historical 90° bend angle data revealed points that were consistently stable about the centerline on the control chart, then an investigation of recent process or raw material changes might be considered.

Estimating the Process Average

Estimates of the process average should be calculated separately for each characteristic or part on short run $\overline{X}$ and range™ charts. In this case, estimates of the process average should be calculated separately for each different spring-back angle. Calculation 17.1 shows the calculation for the overall average of the 30° spring-back measurements.

$$\overline{\overline{X}}_{30^\circ} = \frac{\sum \overline{X}_{30^\circ}}{k_{30^\circ}} = \frac{92.44}{11} = 8.40$$

Calculation 17.1. Estimate of the process average for 30° spring-back angles.

Estimating σ

Estimates of σ are also calculated separately for each characteristic or location represented on short run $\overline{X}$ and range™ charts. In this case, estimates of the process standard deviation should be calculated for each different spring-back angle.

$$\overline{R}_{30^\circ} = \frac{\sum R_{30^\circ}}{k_{30^\circ}} = \frac{4.73}{11} = 0.43$$

Calculation 17.2. Calculation of the average moving range for 30° spring back-angles (to be used in estimating the standard deviation).

$$\hat{\sigma}_{30^\circ} = \frac{\overline{R}_{30^\circ}}{d_2} = \frac{0.43}{2.326} = 0.18$$

Calculation 17.3. Estimate of the process standard deviation for the 30° spring-back angles.

Note: To ensure reliable estimates, k needs to be at least 20. In this example, k is only 11. Therefore, the estimates here and in Table 17.7 should be used only as references.

Calculating Process Capability and Performance Ratios

$$C_{p30^\circ} = \frac{\text{USL}_{30^\circ} - \text{LSL}_{30^\circ}}{6\hat{\sigma}_{30^\circ}} = \frac{8.7 - 7.7}{6(0.18)} = \frac{1.0}{1.08} = 0.93$$

Calculation 17.4. C_p calculation for the 30° bend angle spring-back.

$$C_{pk_u 30^\circ} = \frac{\text{USL}_{30^\circ} - \overline{\overline{X}}_{30^\circ}}{3\hat{\sigma}_{30^\circ}} = \frac{8.7 - 8.4}{3(0.18)} = \frac{0.3}{0.54} = 0.56$$

Calculation 17.5. $C_{pk\,\text{upper}}$ calculation for the 30° bend angle spring-back.

$$C_{pk_l 30^\circ} = \frac{\overline{\overline{X}}_{30^\circ} - \text{LSL}_{30^\circ}}{3\hat{\sigma}_{30^\circ}} = \frac{8.4 - 7.7}{3(0.18)} = \frac{0.7}{0.54} = 1.30$$

Calculation 17.6. $C_{pk\,\text{lower}}$ calculation for the 30° bend angle spring-back.

Short Run $\overline{X}$ *and Range™ Chart Advantages*

- Graphically illustrates the variation of multiple product or process characteristics on the same chart.
- Characteristics from different parts with different means, different standard deviations, and different units of measure can be analyzed on the same chart.
- Pinpoints the characteristics that are in need of the most attention.
- Separates variation due to changes in average from variation due to changes in the standard deviation.
- Separates process variation from product-specific variation.

Short Run $\overline{X}$ *and Range™ Chart Disadvantages*

- The use of negative numbers and unitless ratios may be confusing at first.
- $\overline{\overline{X}}$, $\overline{R}$, and the estimate of σ must be calculated separately for each characteristic on the chart.
- Proper chart analysis requires knowledge of how target values were derived.

An Additional Comment About the Case

The process capability and performance ratio calculations for the 45° and 90° bend angle spring-back are shown in Table 17.7.

Table 17.7. C_p and C_{pk} calculations for 45° and 90° bend angle spring-back characteristics.

45°	90°
$\overline{X}_{45°} = 3.98$	$\overline{X}_{90°} = 1.20$
$\overline{R}_{45°} = 0.28$	$\overline{R}_{90°} = 0.05$
$\hat{\sigma}_{45°} = 0.12$	$\hat{\sigma}_{90°} = 0.02$
$C_{p\,45°} = 1.37$	$C_{p\,90°} = 7.27$
$C_{pk_u\,45°} = 1.42$	$C_{pk_u\,90°} = 8.72$
$C_{pk_l\,45°} = 1.31$	$C_{pk_l\,90°} = 5.82$

Chapter 18

Short Run $\overline{X}$ and s Chart

Decision Tree Section

Table 18.1. Control chart decision tree.

Number of characteristics or locations on the same chart?	Subgroup size?	Different averages on the same chart?	Different standard deviations on the same chart?	Use this chart
1	≥10	Yes	Yes	Short run $\overline{X}$ and s

Description

Short Run $\overline{X}$ *Chart*

The short run $\overline{X}$ chart is used to monitor and detect changes in the averages among multiple characteristics of any type. The characteristics may have different nominals, different units of measure, and different standard deviations. However, all characteristics on the chart should be related enough to warrant analyzing them together. The plot points are coded by subtracting from each subgroup average its respective target $\overline{\overline{X}}$ (usually the engineering nominal value) and then dividing by its target $\bar{s}$. Each characteristic on the chart has its own unique target $\overline{\overline{X}}$ and target $\bar{s}$. For each characteristic, the number of measurements taken in a subgroup is typically 10 or more.

Short Run s Chart

The short run s chart is used to monitor and detect changes in the standard deviations among multiple types of measured characteristics. The plot points are coded by dividing each subgroup's sample standard deviation (s) by its respective target $\bar{s}$.

Subgroup Assumptions

- Independent measurements
- Constant sample size

Calculating Plot Points

Table 18.2. Plot point formulas for short run $\overline{X}$ and s chart.

Chart	Plot point	Plot point formula
Short run $\overline{X}$	Coded $\overline{X}$	$\dfrac{\overline{X} - \text{Target } \overline{\overline{X}}}{\text{Target } \bar{s}}$
Short run s	Coded s	$\dfrac{s}{\text{Target } \bar{s}}$

Calculating Centerlines

Table 18.3. Short run $\overline{X}$ and s chart centerlines.

Chart	Centerline	Centerline formula
Short run $\overline{X}$	0	0
Short run s	1	1

Calculating Control Limits

Table 18.4. Control limits for the short run $\overline{X}$ and s chart.

Chart	Upper control limit	Lower control limit
Short run $\overline{X}$	$+A_3$	$-A_3$
Short run s	B_4	B_3

Example

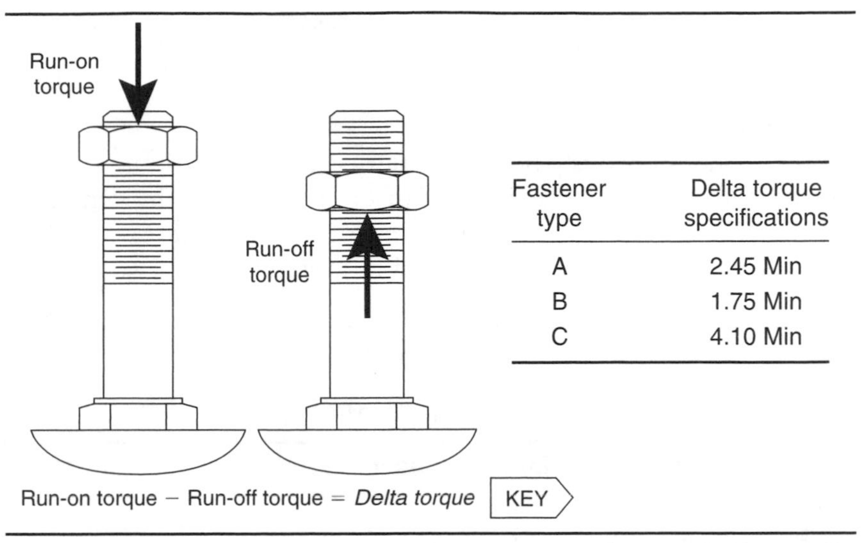

Fastener type	Delta torque specifications
A	2.45 Min
B	1.75 Min
C	4.10 Min

Figure 18.1. Delta torque is a performance key characteristic on self-locking fastener systems.

Case Description

Torque is tested on self-locking nuts using precision stud standards and production nuts. During production, the nuts are slightly deformed so that the threads create an interference or locking fit with the stud. The run-on torque is the average prevailing torque while turning the nut on the stud seven clockwise revolutions. The run-off torque is the maximum force it takes to turn the nut back off the stud one counterclockwise revolution. The delta torque is the run-on torque minus the run-off torque. Each fastening system has its own minimum delta torque requirements and the standard deviations are expected to vary from system to system.

Sampling Strategy

Torque tests are performed for each batch of locking nuts. Ten samples are tested from each batch. To monitor the delta torque consistency, regardless of the nut/bolt locking system, a short run $\overline{X}$ and *s* chart is selected. This is the appropriate chart because the subgroup sizes are large and the standard deviations are different from system to system.

Target Values

Before a short run chart can be used, target values must first be defined.

Locking System A

System A has previously been maintained using traditional $\overline{X}$ and s charts. On the most recent set of in-control charts, the centerline on the $\overline{X}$ chart was 2.920. The centerline on the s chart was 0.089. Therefore, these centerlines are used as target values for system A.

$$\text{Target } \overline{\overline{X}}_A = 2.920$$

$$\text{Target } \bar{s}_A = 0.089$$

Figure 18.2. Target values for locking system A.

Locking System B

The consistency of locking system B has never been evaluated with a control chart. However, quality assurance personnel have taken 28 delta torque measurements at some time in the past. Equation 15.14 was used to convert the sample standard deviation from those 28 measurements into the targets found in Figure 18.3.

$$\text{Target } \overline{\overline{X}}_B = 2.330$$

$$\text{Target } \bar{s}_B = 0.121$$

Figure 18.3. Target values for locking system B.

Locking System C

Like system A, locking system C has previously been evaluated using traditional $\overline{X}$ and s charts. On the most recent set of in-control charts, the centerline on the $\overline{X}$ chart was 5.125. The centerline on the s chart was 0.337. Therefore, these centerlines are used as target values for system C (see Figure 18.4).

$$\text{Target } \overline{\overline{X}}_C = 5.125$$

$$\text{Target } \bar{s}_C = 0.337$$

Figure 18.4. Target values for locking system C.

Data Collection Sheet

Table 18.5. Delta torque data sheet and plot point calculations.

Locking system	A	A	A	A	B	B	C	C
Subgroup number	1	2	3	4	5	6	7	8
1	2.780	2.795	2.747	2.630	2.365	2.649	5.390	5.389
2	2.669	2.787	2.823	2.911	2.385	2.595	4.968	5.387
3	2.944	2.841	2.903	2.910	2.683	2.492	5.205	5.265
4	2.960	2.567	2.849	2.651	2.333	2.603	5.742	5.755
5	2.637	2.868	2.852	2.969	2.232	2.489	5.460	5.121
6	2.598	2.938	2.821	2.779	2.667	2.261	5.376	4.534
7	2.906	2.897	2.807	2.832	2.445	2.657	5.337	5.068
8	2.729	2.864	2.863	2.840	2.415	2.556	4.526	4.939
9	2.738	2.896	2.871	2.809	2.389	2.131	4.653	5.164
10	2.912	2.887	2.862	2.757	2.344	2.466	5.458	5.011
$\bar{X}$	2.787	2.834	2.840	2.809	2.426	2.490	5.212	5.163
Target $\bar{\bar{X}}$	2.920	2.920	2.920	2.920	2.330	2.330	5.125	5.125
$\bar{X}$ − Target $\bar{\bar{X}}$	−0.133	−0.086	−0.080	−0.111	0.096	0.160	0.087	0.038
Target $\bar{s}$	0.089	0.089	0.089	0.089	0.121	0.121	0.337	0.337
$\frac{\bar{X} - \text{Target } \bar{\bar{X}}}{\text{Target } \bar{s}}$	−1.491	−0.966	−0.901	−1.249	0.792	1.321	0.257	0.114
s	0.134	0.105	0.043	0.110	0.143	0.171	0.383	0.323
s/Target $\bar{s}$	1.509	1.180	0.482	1.233	1.183	1.413	1.136	0.960
Locking system	C	A	A	A	C	C	B	B
Subgroup number	9	10	11	12	13	14	15	16
1	5.083	2.823	2.866	2.801	4.832	6.120	2.253	2.248
2	4.869	2.754	2.962	2.790	5.392	4.904	2.292	2.401
3	4.802	2.775	2.685	2.919	5.225	5.477	2.340	2.059
4	5.193	2.887	2.823	2.814	4.964	4.646	2.243	2.142
5	5.063	2.883	2.847	2.832	4.828	5.730	2.395	2.352
6	4.257	2.989	2.787	2.600	5.896	4.391	2.376	2.447
7	5.703	2.959	2.943	2.977	5.220	5.400	2.351	2.572
8	4.475	2.811	2.871	2.817	5.691	5.051	2.681	2.338
9	4.489	2.885	2.767	2.718	5.510	5.745	2.497	2.248
10	4.915	2.801	2.807	2.918	4.911	4.424	2.548	2.479
$\bar{X}$	4.885	2.857	2.836	2.819	5.247	5.189	2.398	2.329
Target $\bar{\bar{X}}$	5.125	2.920	2.920	2.920	5.125	5.125	2.330	2.330
$\bar{X}$ − Target $\bar{\bar{X}}$	−0.240	−0.063	−0.084	−0.101	0.122	0.064	0.068	−0.001
Target $\bar{s}$	0.337	0.089	0.089	0.089	0.337	0.337	0.121	0.121
$\frac{\bar{X} - \text{Target } \bar{\bar{X}}}{\text{Target } \bar{s}}$	−0.712	−0.711	−0.946	−1.139	0.362	0.189	0.559	−0.012
s	0.417	0.077	0.082	0.108	0.373	0.598	0.139	0.157
s/Target $\bar{s}$	1.236	0.870	0.922	1.210	1.106	1.773	1.152	1.299

Table 18.5. Cont.

Locking system	B	B	B	B
Subgroup number	17	18	19	20
1	2.507	2.360	2.062	2.278
2	2.611	2.418	2.841	2.299
3	2.572	2.102	2.270	2.432
4	2.426	2.247	2.542	2.512
5	2.211	2.305	2.425	2.586
6	2.533	2.227	2.621	2.251
7	2.303	2.388	2.491	2.694
8	2.614	2.533	2.709	2.511
9	2.350	2.235	2.458	2.459
10	2.526	2.686	2.253	2.226
$\overline{X}$	2.465	2.350	2.467	2.425
Target $\overline{\overline{X}}$	2.330	2.330	2.330	2.330
$\overline{X}$ − Target $\overline{\overline{X}}$	0.135	0.020	0.137	0.095
Target $\overline{s}$	0.121	0.121	0.121	0.121
$\dfrac{\overline{X} - \text{Target } \overline{\overline{X}}}{\text{Target } \overline{s}}$	1.118	0.166	1.134	0.783
s	0.138	0.168	0.231	0.157
s/Target $\overline{s}$	1.138	1.391	1.906	1.297

Short Run $\overline{X}$ and s Chart

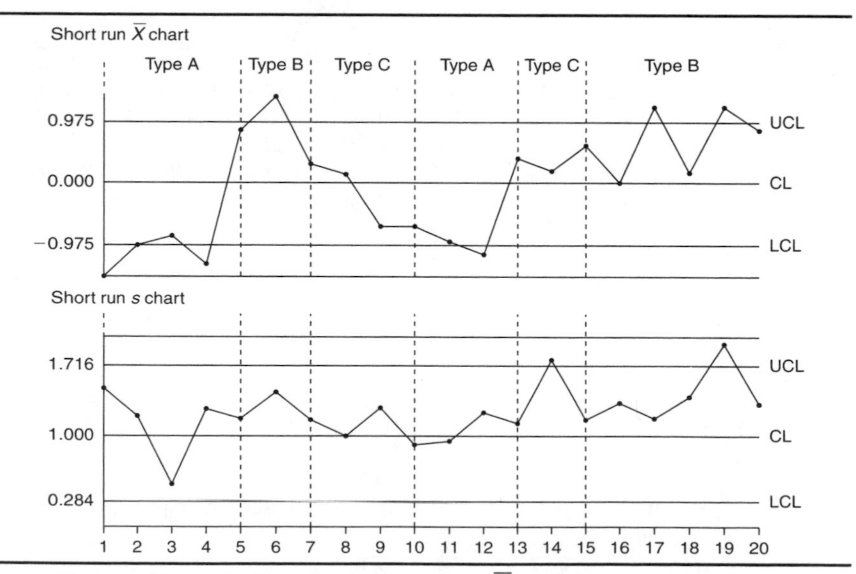

Figure 18.5. Delta torque short run $\overline{X}$ and s control charts for locking systems A, B, and C.

Chart Interpretation

Short run s *chart:* If evaluating product-specific variation, locking system A's delta torque seems to be behaving randomly. All eight of system B's plot points fall above the centerline with one of them falling above the UCL. System C's delta torque favors the high side with one plot point beyond the UCL. Overall, the process reveals a run of 9 plot points above the centerline that occur across three product lines (subgroups 13 through 20).

Short run $\overline{X}$ *chart:* All seven of system A's plot points fall below the centerline with three of them falling below the LCL. Seven of system B's eight plot points are situated above the centerline with three above the UCL. System C appears to be behaving randomly. Looking at patterns across locking systems, there is a gradual decrease in the average from plot point 6 through 12. Also, it looks as though the average has shifted higher between plot points 13 and 20.

Recommendations

> **Note:** Plot point patterns above and below the centerlines and beyond the control limits are present, but the action to take depends entirely on how the target values were estimated.

Locking System A

Short run s *chart:* The target $\bar{s}$ came from past control charts, therefore, the fact that the plot points are behaving randomly indicates that the standard deviation has not changed since data were last recorded.

Short run $\overline{X}$ *chart:* The target $\overline{\overline{X}}$ came from past charts, therefore, the run below the centerline indicates the delta torque has decreased since data were last recorded. This is an assignable cause and should be investigated. If the shift is found to be desirable, deliberate, and permanent, the target $\overline{\overline{X}}$ should be recalculated based on system A's current overall average. If the shift is found to be an unwanted condition, do not recalculate target $\overline{\overline{X}}$. Instead, eliminate the cause of the downward shift.

Locking System B

Short run s *chart:* The target $\bar{s}$ came from past quality assurance records. The run above the centerline, therefore, indicates that the standard deviation has significantly increased since data were last recorded. This may be an assignable cause and should be investigated. If the shift is found to be an unwanted condition, do not recalculate target $\bar{s}$. Instead, eliminate the cause of the increased variability.

Short run $\overline{X}$ *chart:* The target $\overline{\overline{X}}$ came from quality assurance records, therefore, the run above the centerline indicates the delta torque has increased since data were last recorded. This may be an assignable cause and should be investigated. If this significant increase in delta torque is desirable, then the target $\overline{\overline{X}}$ should be recalculated based on system B's current overall average. If the shift is unwanted, do not recalculate target $\overline{\overline{X}}$. Instead, eliminate the assignable cause for the increase in the delta torque average.

Locking System C

Short run s *chart:* Because the target $\bar{s}$ was based on the centerline from an older, in-control *s* chart, the run above the centerline indicates that the process standard deviation has increased significantly since the last time the system C product was manufactured. This should be treated as an assignable cause because the target $\bar{s}$ is based upon actual data. If the increase in standard deviation for system C is expected to be a permanent change, then the target $\bar{s}$ should be recalculated based on the current overall average standard deviation (see Calculation 18.1). Otherwise, if the assignable cause is to be removed to reduce the current amount of variation, the old target $\bar{s}$ should be saved to represent the current expected level of variability.

$$\text{Updated target } \bar{s} = \frac{\sum s_c}{k_c}$$

$$= \frac{0.383 + 0.323 + 0.417 + 0.373 + 0.598}{5}$$

$$= \frac{2.094}{5}$$

$$= 0.419$$

Calculation 18.1. Recalculating locking system C's target $\bar{s}$ based on current data from control chart. This is done only if the change in variability is expected to be a permanent one.

Short run $\overline{X}$ *chart:* The target $\overline{\overline{X}}$ has been obtained from a recent in-control chart, and the plot points are behaving randomly. This indicates that the initial target $\overline{\overline{X}}$ was a good estimator of the actual delta torque. There is no need to recalculate system C's target $\overline{\overline{X}}$.

Estimating the Process Average

Estimates of the process average should be calculated separately for each characteristic or part on short run $\overline{X}$ and s charts. In this case, estimates of the process average should be calculated separately for each different locking system. Calculation 18.2 shows the calculation for the estimate of the overall average of locking system B.

$$\overline{\overline{X}}_B = \frac{\sum \overline{X}_B}{k_B}$$

$$= \frac{2.426 + 2.490 + 2.398 + 2.329 + 2.465 + 2.350 + 2.467 + 2.425}{8}$$

$$= \frac{19.350}{8}$$

$$= 2.419$$

Calculation 18.2. Estimate of the process average for locking system B.

Estimating σ

Estimates of σ are also calculated separately for each characteristic or location represented on short run $\overline{X}$ and s charts. In this case, estimates of the process standard deviation should be calculated for each different locking system. Estimates of the process standard deviation for locking system B are found in Calculation 18.3.

$$\bar{s}_B = \frac{\sum s_B}{k_B}$$

$$= \frac{0.143 + 0.171 + 0.139 + 0.157 + 0.138 + 0.168 + 0.231 + 0.157}{8}$$

$$= \frac{1.304}{8}$$

$$= 0.163$$

Calculation 18.3. Calculation of $\bar{s}$ for locking system B based on current data from the short run s control chart.

$$\hat{\sigma}_B = \frac{\bar{s}_B}{c_4} = \frac{0.163}{0.9727} = 0.168$$

Calculation 18.4. Calculation of the estimate of the process standard deviation for locking system B.

Note: To ensure reliable estimates, k needs to be at least 20. In this example, k is only 8. Therefore, the estimates here and in Table 18.6 are used for illustration purposes only.

Calculating Process Capability and Performance Ratios

The C_{pk_l} calculation for locking system B is shown in Calculation 18.5. Because there is only a minimum specification, no C_p or C_{pk_u} value is calculated for locking system B.

$$C_{pk_l B} = \frac{\overline{\overline{X}}_B - LSL_B}{3\hat{\sigma}_B} = \frac{2.419 - 1.75}{3(0.168)} = \frac{0.669}{0.504} = 1.33$$

Calculation 18.5. C_{pk_l} calculation for fastener system B delta torque.

Short Run $\overline{X}$ *and* s *Chart Advantages*

- Graphically illustrates the variation of multiple products with different nominals, different standard deviations, and even different units of measure all on the same chart.
- Separates sources of process variability from sources of product variability.
- Due to the large sample sizes, the short run $\overline{X}$ chart is sensitive to small changes in the process average.
- Summarizes large amounts of data.

Short Run $\overline{X}$ *and* s *Chart Disadvantages*

- Requires software to effectively handle large amounts of data
- The use of negative numbers and unitless ratios may be confusing at first.
- $\overline{\overline{X}}$, $\bar{s}$, and process standard deviation estimates must be calculated separately for each characteristic represented on the chart.

Additional Comments About the Case

- The process capability and performance ratio calculations for locking systems A and C are found in Table 18.6.
- Summary statistics and C_{pk_l} values for systems A and C are based on the actual data from the data collection sheet (Table 18.5). In addition, no C_p or C_{pk_u} values are found in Table 18.6 because the locking systems all have one-sided specifications.

Table 18.6. Additional summary statistics and process performance ratios for locking systems A and C.

Locking system A	Locking system C
$\overline{\overline{X}}_A = 2.826$	$\overline{\overline{X}}_C = 5.139$
$\overline{s}_A = 0.094$	$\overline{s}_C = 0.419$
$\hat{\sigma}_A = 0.097$	$\hat{\sigma}_C = 0.431$
$C_{pk_l\,A} = 1.29$	$C_{pk_l\,C} = 0.80$

Chapter 19

Charts for Multiple Characteristics—The Group Charts

Almost every SPC user has encountered situations where a single part or process had several characteristics that required control chart evaluation. For example, one might need to measure the gap between an automobile's door and its body panels at multiple locations around the door's perimeter. The interest might be in evaluating the consistency of the gap from location to location on one door and/or from one car to the next. If data were taken at five locations around the door perimeter, five separate control charts would need to be developed—one for each of the five gap locations of interest. Of course, this translates into added complexity, work, and interpretation for those who want to evaluate gap consistency.

However, wouldn't it be nice to evaluate multiple characteristics on only one chart? If one could, record keeping, data collection, and interpretation would be simplified. These benefits can be achieved if one uses group charts. In Eugene L. Grant's book, *Statistical Quality Control,* the author states that group charts are "an ingenious method for combining a number of subgroups from different sources in a single simplified chart . . . the technique used is broadly applicable."[1] Group charts have been developed specifically to allow SPC practitioners to evaluate multiple characteristics on a single chart.

All group charts share common features. Rather than repeating these commonalties in each group chart chapter, they will be covered in this overview.

Group Charts Are Based on Traditional Control Charts

The three traditional—or core—variables control charts are

1. Individual X and moving range chart
2. $\overline{X}$ and range chart
3. $\overline{X}$ and s chart

All three of the core variables control charts are used to monitor the stability of a process' central tendency (average) and the variation about that average (standard deviation). All are single-variable control charts. That is, only one characteristic can be monitored on the chart. The group charts described in this section are based upon the core traditional charts just listed.

Group Chart Defined

The family of group charts described in this chapter is found on the bottom half of the control chart decision tree (see Figure 6.1). Group charts are based on the same criteria as their traditional counterparts, with one exception: several characteristics, parameters, or process streams can be combined on the same chart.

Data can be grouped whenever it would take more than one single-variable chart to fully describe the behavior of a process or component at any given point in time. Group charts are ideal in situations where measurements from several locations are required to ensure uniformity. Examples where group charts might be used include the monitoring of

- Consistent fit and fair at several locations between a leading edge and a wing surface
- Roundness of a hole
- Within-part flange thickness consistency
- Temperature and pressure consistency at several locations in an autoclave

Group charts are also recommended when monitoring processes where x and y or x, y, and z axes are simultaneously monitored while manufacturing products. Examples include geometric tolerances and three-dimension spatial locations from coordinate measurement machines.

All group charts are similar in construction and interpretation. The only distinction is in the calculation of the plot points. For example, group $\overline{X}$ chart plot points represent actual subgroup averages, but group target $\overline{X}$ chart plot points represent the difference between subgroup averages and their specified target values. In other words, the distinction among the different group charts is the same distinction among their single-variable control chart counterparts.

How to Construct a Group Chart

1. Subgroups of data are logically combined into a group.
2. Plot points from each subgroup are calculated.
3. The largest (MAX) and smallest (MIN) plot points within each group are plotted on the appropriate group chart.
4. Once plotted, the MAX points are connected together with a line as are the MIN points. The result is two lines of connected data points on each chart.
5. MAX and MIN plot points are labeled with their respective locations from the part or process.

Taper Example

To illustrate the mechanics of constructing a group chart, consider a simple machined shaft (see Figure 19.1). To monitor uniformity of the diameter, measurements are taken of the outside diameter (OD) at locations A, B, and C.

The number of characteristics being monitored is three (locations A, B, and C), the subgroup size is four, and it is assumed the standard deviation is similar at all three locations. A group $\overline{X}$ and R chart will be used to monitor this part. (The control chart decision tree—Figure 6.1—was consulted when making this decision.)

The data in Table 19.1 represent samples from five groups. Each group represents four shafts, each measured at locations A, B, and

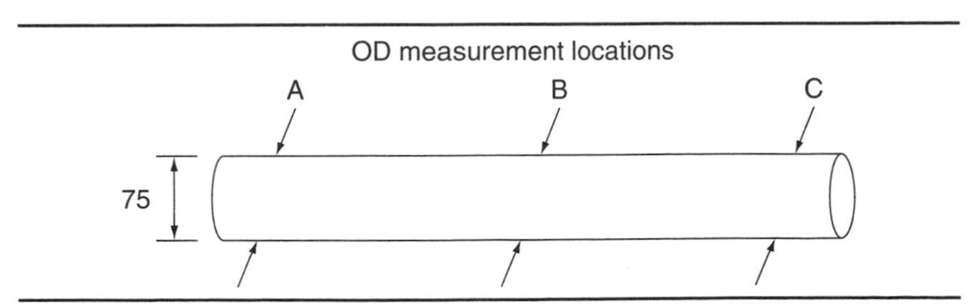

Figure 19.1. Machined shaft OD measured at three locations.

Table 19.1. Group chart data collection sheet for the machined shaft example.

Group	1			2			3			4			5		
Location	A	B	C	A	B	C	A	B	C	A	B	C	A	B	C
Part 1	75	72	74	77	77	75	71	75	72	76	73	74	73	77	76
Part 2	74	74	74	75	70	75	76	72	72	76	76	72	76	76	74
Part 3	73	75	73	76	75	74	75	74	74	75	76	75	76	76	74
Part 4	72	77	71	76	72	74	78	77	73	75	75	75	75	74	74
Average	73.5	74.5	73	76	73.5	74.5	75	74.5	72.75	75.5	75	74	75	75.75	74.5
Range	3	5	3	2	7	1	7	5	2	1	3	3	3	3	2

C. The average and range at each location has been calculated with MAX and MIN values circled within each group. In groups 1, 4, and 5, ties occurred in the ranges.

Only the circled MAX and MIN values in Table 19.1 are plotted on the group chart. Table 19.2 displays a listing of the circled plot points and corresponding location letters from each group found in Table 19.1. Regardless of the number of locations measured in each group, only four plot points result from each group: MAX average, MIN average, MAX range, and MIN range. If there had been five locations measured on the shaft instead of three, there would still only be four plot points for each group.

The MAX and MIN averages are plotted on the group $\overline{X}$ chart and the MAX and MIN ranges are plotted on the group range chart. MAX plot points are connected to each other and MIN plot points are connected to each other. Each plot point is labeled with an A, B, or C to represent its location on the shaft (see Figure 19.2).

Table 19.2. Summary plot point group chart data from Table 19.1.

Group	1	2	3	4	5
MAX average	74.5	76.0	75.0	75.6	75.75
Location	B	A	A	A	B
MIN average	73.0	73.5	72.75	74.0	74.5
Location	C	B	C	C	C
MAX range	5	7	7	3	3
Location	B	B	A	BC	AB
MIN range	3	1	2	1	2
Location	AC	C	C	A	C

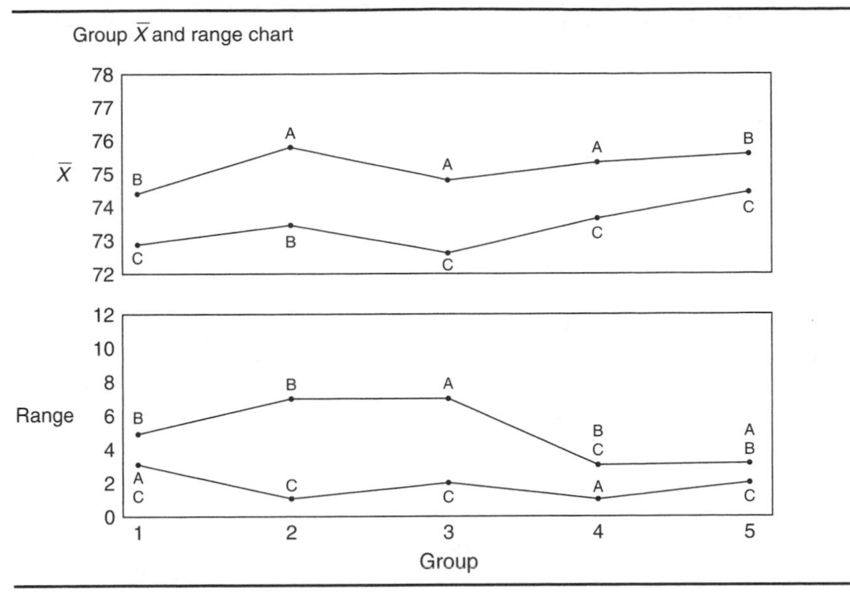

Figure 19.2. Group $\overline{X}$ and range chart for the machined shaft example.

Interpreting the Shaft Taper Example Group Chart

On the group $\overline{X}$ chart (Figure 19.2), location C appears four out of five times as the MIN plot point. This indicates that the diameter at location C is consistently smaller than diameters A or B.

Location C also appears on the group range chart (Figure 19.2) as the MIN plot point four out of five times. This would indicate that the piece-to-piece variation at diameter C is consistently less than at locations A or B. Also on the group range chart, location B appears four out of five times as the MAX plot point. This indicates that location B's standard deviation is greater than locations A or C.

When more than one location within a grouping has the same average or range, the plot point may be labeled with the locations that tied. On the group range chart, because A and C both had the same MIN value in group 1, the plot point is labeled "AC."

General Interpretation of a Group Chart

When interpreting group charts, one should look for consecutive repeating locations appearing in the MAX or MIN position. If this occurs, it is called a *run* and would indicate a consistent difference (assignable cause relative to the other locations) at that location. On a group chart, a run of four or more typically indicates the presence of an assignable cause.

If all locations grouped on a chart were behaving randomly relative to each other, the MAX and MIN plot point labels would reveal no runs.

The gap between the MAX and MIN lines on a group $\overline{X}$ chart represents the average within-piece variation. A small gap represents small within-piece variation. If there is no within-piece variation, the MAX and MIN lines converge. If the MAX and MIN lines are parallel, the characteristics in the group are dependent upon each other. That is, there is a source of variation present that affects all characteristics on the chart in the same manner.

When to Group Data Together

Data should be grouped only if there is a logical reason for doing so. The most common reason to group data is when the same dimension is measured at several locations across a part. The previous shaft example is an example of this. Another example would be flange thickness. Thickness measurement data from several locations across the part could be combined into a single group chart.

Unrelated data should not be grouped together. Going back to the shaft example, if an OD is created using a lathe and then the shaft is gun-drilled to produce an inside diameter, the OD measurements should not be grouped with the inside diameter measurements. Even though they are both diameter measurements used to detect taper, the processes used to make the diameters are very different. The data from the lathe and the data from the gun-drill should be analyzed on separate group charts. However, it would be appropriate to plot the resulting wall thickness from several locations on the same part with a group chart.

Sometimes the decision of when to group data is not so clear. The more the data or characteristics have in common, the better. If dissimilar characteristics are grouped, proper analysis of the chart can be difficult. Also, the number of characteristics or measurement locations on a single group chart should be held to five or less. If more than five are used, there is a higher likelihood of missing a trend—it could be hidden between the MAX and MIN plot points.

Group Charts and Control Limits

There are two distinct uses for group charts.

1. They can be used for process control. As such, control limits are appropriate.
2. They can be used strictly as an exploratory analysis tool. In this case, control limits are not necessary.

To use a group chart for process control, one groups data streams from multiple processes onto a single control chart. A data stream is a collection of time-ordered data that represents the output of a single process. By grouping data from similar but independent processes (or coding the data so that dissimilar independent data streams can be analyzed together) one can use a single set of control limits to represent all processes on the chart. This approach is most commonly used where multiple process lines are simultaneously used to produce identical or similar product. Using group charts in this situation not only cuts down on the number of control charts, it also helps focus on processes whose output begins to change significantly compared to other processes. When all data streams on the group chart are independent, control limits are acceptable.

In this text, however, we chose to describe group charts as exploratory analysis tools. In many cases, one is not necessarily interested in process control, but rather the time-ordered comparison among multiple data streams regardless of part numbers or data stream interdependence. Since examples presented in the following group chart chapters combine dependent data streams and are strictly exploratory in nature, no control limits are used; therefore we do not refer to them as control charts.

Note

1. Eugene L. Grant, *Statistical Quality Control,* 3d ed. (New York: McGraw-Hill, 1964), 161.

Chapter 20

Group Individual *X* and Moving Range Chart

Decision Tree Section

Table 20.1. Control chart decision tree.

Number of characteristics or locations on the same chart?	Subgroup size?	Different averages on the same chart?	Different standard deviations on the same chart?	Use this chart
>1	1	No	No	Group *IX–MR* chart

Description

The Group Individual X *Chart*

The *group individual X* chart, also called a *group IX chart,* is used to monitor and detect changes in the average among multiple measurement locations of an identical type of characteristic. Individual measurements from these multiple locations are combined into a group. The plot points represent the maximum and minimum individual measurements from a grouping. The number of measurements taken from each location in a subgroup is one. For best results, the number of measurement locations in a group should not exceed five.

The Group Moving Range Chart

The group moving range chart, also called a *group MR chart,* is used to monitor and detect changes in the standard deviation among multiple measurement locations of an identical type of characteristic. Moving ranges from multiple locations are combined into a group. The plot points represent the maximum and minimum moving ranges between consecutive groups. Moving ranges are not the same as standard deviations, but they can be used to estimate standard deviation.

Subgroup Assumptions

- Independent measurements
- Sample size of one
- Similar characteristics
- One unit of measure

Calculating Plot Points

Table 20.2. Plot point formulas for the group *IX* and *MR* chart.

Chart	Plot point	Plot point formula
Group individual *X*	MAX and MIN *IX*	Individual measurement
Group moving range	MAX and MIN *MR*	Moving range between two consecutive individual measurements from the same location

Calculating Centerlines

Table 20.3. Group *IX* and *MR* chart centerline formulas.

Chart	Centerline	Centerline formula
Group individual *X*	$\overline{IX}$	$\frac{\sum IX}{k}$
Group moving range	$\overline{MR}$	$\frac{\sum MR}{k-1}$

Calculating Control Limits

Table 20.4. Group *IX* and *MR* chart control limits.

Chart	Upper control limit	Lower control limit
Group individual *X*	none	none
Group moving range	none	none

Example

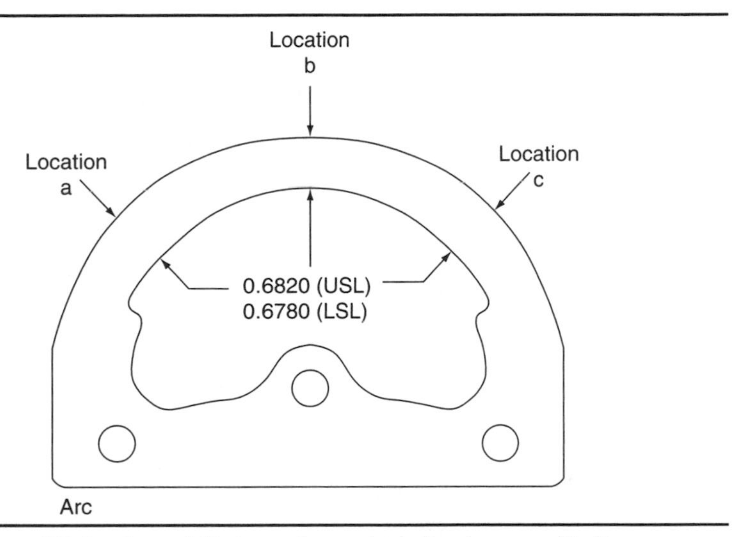

Figure 20.1. Arc width key characteristic shown with three measurement locations and upper and lower specifications.

Case Description

The arc shown in Figure 20.1 is a sheet metal stamping. It serves as a guide for a tractor throttle control. For the throttle assembly to function correctly, the arc width must be uniform and within specification. If the width is too large, the assembly binds, if it is too small, the assembly will not lock into position. To monitor arc width uniformity, measurements are taken at three locations, a, b, and c. The quality department wants to use a chart that will examine all three locations simultaneously.

Sampling Strategy

Because the same characteristic is being measured at three different locations on the part and there is an interest in evaluating them all on one chart, a group individual *X* and moving range chart is used.

Data Collection Sheet

Table 20.5. Group *IX–MR* chart data collection sheet. MAX and MIN plot points are shown in bold.

Group	1			2			3		
Location	a	b	c	a	b	c	a	b	c
IX	***0.6813***	***0.6790***	0.6792	***0.6813***	***0.6785***	0.6791	***0.6811***	0.6799	***0.6793***
MR	—	—	—	***0.0000***	***0.0005***	0.0001	***0.0002***	***0.0014***	***0.0002***
Group	4			5			6		
Location	a	b	c	a	b	c	a	b	c
IX	***0.6810***	***0.6792***	0.6795	***0.6816***	0.6793	***0.6792***	***0.6811***	***0.6789***	0.6795
MR	***0.0001***	***0.0007***	0.0002	***0.0006***	***0.0001***	0.0003	***0.0005***	0.0004	***0.0003***
Group	7			8			9		
Location	a	b	c	a	b	c	a	b	c
IX	***0.6810***	0.6797	***0.6793***	***0.6812***	0.6802	***0.6795***	***0.6814***	***0.6793***	0.6795
MR	***0.0001***	***0.0008***	0.0002	***0.0002***	***0.0005***	***0.0002***	0.0002	***0.0009***	***0.0000***

Plot Point Calculations

The Group Individual X *Chart*

No calculations are required for the group *IX*. The MAX and MIN plot points are picked from the individual measurements. For example, in group 1, the largest (MAX) arc width is 0.6813 at location a. The smallest (MIN) width is 0.6790 at location b.

The Group Moving Range Chart

The moving range is calculated by taking the absolute difference between individual measurements at the same location from two consecutive groups. For example, location a in group 2 is 0.6813 and location a in group 3 is 0.6811, so the moving range between the two groups is 0.0002. The moving range at location a between groups 1 and 2 is 0.0000 because the arc width is 0.6813 in both groups for the a location. The same calculations are performed for locations b and c.

> **Note:** There is no moving range for group 1 because no previous measurements exist.

Group IX *and* MR *Chart Plot Points*

Table 20.6. Group *IX* and *MR* chart plot point summary.

Group	1	2	3	4	5	6	7	8	9
MAX *IX*	0.6813	0.6813	0.6811	0.6810	0.6816	0.6811	0.6810	0.6812	0.6814
Location	a	a	a	a	a	a	a	a	a
MIN *IX*	0.6790	0.6785	0.6793	0.6792	0.6792	0.6789	0.6793	0.6795	0.6793
Location	b	b	c	b	c	b	c	c	b
MAX *MR*	—	0.0005	0.0014	0.0007	0.0006	0.0005	0.0008	0.0005	0.0009
Location	—	b	b	b	a	a	b	b	b
MIN *MR*	—	0.0000	0.0002	0.0001	0.0001	0.0003	0.0001	0.0002	0.0000
Location	—	a	a, c	a	b	c	a	a, c	c

Group IX *and* MR *Chart*

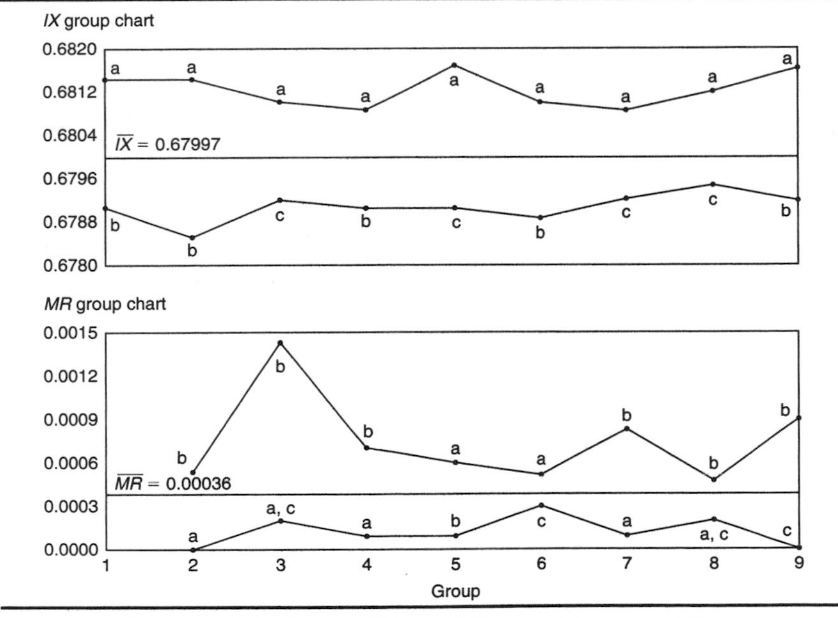

Figure 20.2. Group *IX* and *MR* chart for arc widths.

Chart Interpretation

Group moving range chart: Location b appears in the MAX position six out of eight times. This suggests that location b has the largest standard deviation of all three locations. Location a appears in the MIN position in five of the eight groups. This suggests that the variability at location a may be less than the other two locations.

> **Note:** The centerline ($\overline{MR} = 0.00036$) is the average of all the ranges from the data sheet, not just the average of the MAX and MIN ranges.

Group individual X *Chart:* Location a dominates the MAX position. This means that the arc width at location a is consistently wider than locations b or c. Locations b and c are both found in the MIN position. Even though location c is MIN more often, the raw data show that the individual values for locations b and c are very similar.

The distance between the MAX and MIN lines on the *IX* chart—0.0023 at plot point 1 and 0.0021 at plot point 9—are indicators of the amount of taper across the arc.

> **Note:** The centerline ($\overline{IX} = 0.67997$) is the average of all the individual measurements from all nine groups.

Recommendations

- The consistently larger thickness at location a should be reduced to make the location less prone to binding.
- The variability at location b might be decreased by modifying the tooling to make the arc more rigid at location b when stamping.

This example is typical of what is found in many products with within-piece variation problems. The group chart helps to detect and highlight those consistently high and low values.

Estimating the Process Average

Process average estimates should be performed separately for each characteristic or location on the group chart.

$$\overline{IX_a} = \frac{\sum IX_a}{k_a} = \frac{6.1310}{9} = 0.68122$$

Calculation 20.1. Estimate of the process average for location a.

Estimating σ

Estimates of σ are also calculated separately for each characteristic or location on the group chart. Continuing with location a, see Calculation 20.2.

$$\overline{MR}_a = \frac{\sum MR_a}{k_a - 1} = \frac{0.0019}{8} = 0.00024$$

$$\hat{\sigma}_a = \frac{\overline{MR}}{d_2} = \frac{0.00024}{1.128} = 0.00021$$

Calculation 20.2. Estimate of the process standard deviation for location a.

Note: To ensure reliable estimates, k needs to be at least 20. In this example, k is nine. Therefore, the estimates found here are used only for illustration purposes.

Calculating Process Capability and Performance Ratios

Calculations 20.3 through 20.5 show the process capability and performance calculations for location a.

$$C_{pa} = \frac{\text{USL}_a - \text{LSL}_a}{6\hat{\sigma}} = \frac{0.6820 - 0.6780}{6(0.00021)} = \frac{0.0040}{0.00126} = 3.17$$

Calculation 20.3. C_p for location a.

$$C_{pk_u a} = \frac{\text{USL}_a - \overline{IX}_a}{3\hat{\sigma}} = \frac{0.6820 - 0.68122}{3(0.00021)} = \frac{0.00078}{0.00063} = 1.24$$

Calculation 20.4. C_{pku} for location a.

$$C_{pk_l a} = \frac{\overline{IX}_a - \text{LSL}_a}{3\hat{\sigma}} = \frac{0.68122 - 0.6780}{3(0.00021)} = \frac{0.00322}{0.00063} = 5.11$$

Calculation 20.5. C_{pkl} for location a.

Group IX–MR *Chart Advantages*

- Graphically illustrates the variation of multiple product or process characteristics simultaneously and relative to each other

cess characteristics simultaneously and relative to each other
- Quickly pinpoints the characteristics that are in need of the most attention

Group IX–MR *Chart Disadvantages*

- Not as sensitive to changes in the process average as the group $\overline{X}$ and range chart
- No visibility of the characteristics that fall between the MAX and MIN plot points
- Cannot detect certain out-of-control conditions because the group charts shown here have no control limits

An Additional Comment About the Case

Process capability and performance calculations for locations b and

Table 20.7. Process capability and performance calculations for locations b and c.

Location b	Location c
$\overline{IX}_b = 0.67933$	$\overline{IX}_c = 0.67934$
$\overline{MR}_b = 0.00066$	$\overline{MR}_c = 0.00019$
$\hat{\sigma}_b = 0.00059$	$\hat{\sigma}_c = 0.00017$
$C_{p\,b} = 1.14$	$C_{p\,c} = 4.01$
$C_{pk_u\,b} = 1.52$	$C_{pk_u\,c} = 5.33$
$C_{pk_l\,b} = 0.75$	$C_{pk_l\,c} = 2.69$

Chapter 21

Group $\overline{X}$ and Range Chart

Decision Tree Section

Table 21.1. Control chart decision tree.

Number of characteristics or locations on the same chart?	Subgroup size?	Different averages on the same chart?	Different standard deviations on the same chart?	Use this chart
>1	>1 <10	No	No	Group $\overline{X}$ and range chart

Description

Group $\overline{X}$ Chart

The group $\overline{X}$ chart is used to monitor and detect changes in the averages among multiple measurement locations of an identical characteristic. At each location, several measurements are taken and an average is calculated. Averages for multiple locations are placed into a group. The plot points on the group $\overline{X}$ chart represent the MAX and MIN averages from the group.

For best results, the number of measurement locations in a group should not exceed five. The number of measurements used to calculate an average at each location may range from two to nine, but subgroup sizes are typically between three and five.

Group Range Chart

The group range chart is used to monitor and detect changes in the standard deviation among multiple measurement locations of an identical characteristic. At each location, several measurements are taken and a range value is calculated. Ranges for multiple locations are placed into a group. The plot points on the group range chart represent the MAX and MIN group ranges.

Subgroup Assumptions

- Independent measurements within a subgroup
- Constant sample size
- One characteristic type
- One unit of measure

Calculating Plot Points

Table 21.2. Formulas for calculating plot points on the group $\overline{X}$ and range chart.

Chart	Plot point	Plot point formula
Group $\overline{X}$	MAX and MIN $\overline{X}$	$\frac{\Sigma X}{n}$
Group range	MAX and MIN range	Largest X – Smallest X

Calculating Centerlines

Table 21.3. Formulas for calculating centerlines on the group $\overline{X}$ and range chart.

Chart	Centerline	Centerline formula
Group $\overline{X}$	$\overline{\overline{X}}$	$\frac{\Sigma \overline{X}}{\text{(Number of groups)(Number of characteristics per group)}}$
Group range	$\overline{R}$	$\frac{\Sigma R}{\text{(Number of groups)(Number of characteristics per group)}}$

Calculating Control Limits

Table 21.4. Control limits for the group $\overline{X}$ and range chart.

Chart	Upper control limit	Lower control limit
Group $\overline{X}$	none	none
Group range	none	none

Example

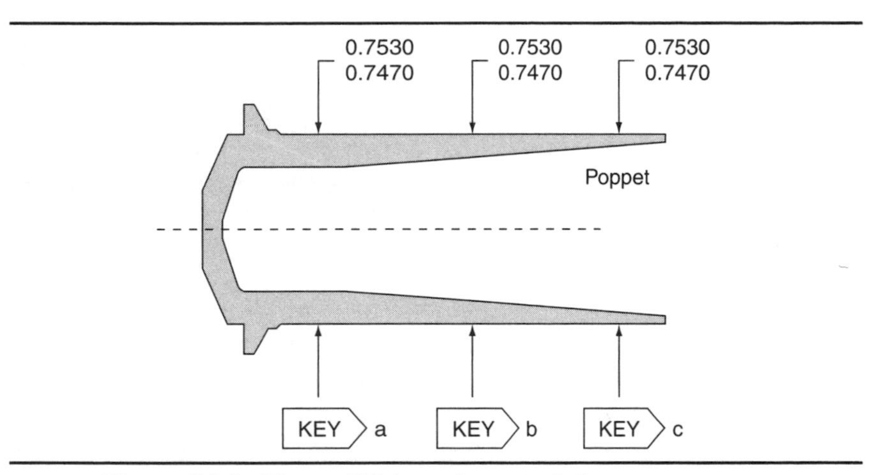

Figure 21.1. Three OD key characteristics on a poppet.

Case Description

A poppet is manufactured on a screw machine. Rejection rates due to inconsistent ODs have been unacceptably high. Therefore, uniformity of the OD is designated as a key characteristic. To check the uniformity, three OD measurements are taken on each poppet at locations a, b, and c. Although the dimensions of the poppet could also be monitored using three separate $\overline{X}$ and range charts—one for each dimension—quality assurance wants to monitor the diameter using only one chart. This is why the group $\overline{X}$ and range chart is selected.

Sampling Strategy

Because the same characteristic is being measured at three different locations on the same part, a group $\overline{X}$ and range chart is selected. Three poppets are measured every 15 minutes.

Data Collection Sheet

Table 21.5. Data collection sheet for the group X and range chart. MAX and MIN plot points for each group are displayed in bold.

Group number	1			2			3		
Location	a	b	c	a	b	c	a	b	c
Sample 1	0.7500	0.7500	0.7470	0.7490	0.7510	0.7480	0.7500	0.7500	0.7480
2	0.7500	0.7510	0.7480	0.7490	0.7500	0.7460	0.7500	0.7490	0.7460
3	0.7500	0.7500	0.7490	0.7500	0.7510	0.7460	0.7500	0.7500	0.7460
$\overline{X}$	0.7500	***0.7503***	***0.7480***	0.7493	***0.7507***	***0.7467***	***0.7500***	0.7497	***0.7467***
Range	***0.0000***	0.0010	***0.0020***	***0.0010***	***0.0010***	***0.0020***	***0.0000***	0.0010	***0.0020***
Group number	4			5			6		
Location	a	b	c	a	b	c	a	b	c
Sample 1	0.7500	0.7510	0.7470	0.7510	0.7510	0.7480	0.7510	0.7500	0.7470
2	0.7510	0.7500	0.7480	0.7500	0.7500	0.7480	0.7500	0.7510	0.7500
3	0.7490	0.7490	0.7500	0.7500	0.7500	0.7480	0.7500	0.7480	0.7480
$\overline{X}$	***0.7500***	***0.7500***	***0.7483***	***0.7503***	***0.7503***	***0.7480***	***0.7503***	0.7497	***0.7483***
Range	***0.0020***	***0.0020***	***0.0030***	***0.0010***	***0.0010***	***0.0000***	***0.0010***	***0.0030***	***0.0030***
Group number	7			8			9		
Location	a	b	c	a	b	c	a	b	c
Sample 1	0.7500	0.7500	0.7480	0.7500	0.7510	0.7480	0.7500	0.7490	0.7470
2	0.7500	0.7500	0.7460	0.7500	0.7480	0.7470	0.7510	0.7500	0.7460
3	0.7500	0.7510	0.7460	0.7500	0.7500	0.7490	0.7500	0.7490	0.7500
$\overline{X}$	0.7500	***0.7503***	***0.7467***	***0.7500***	0.7497	***0.7480***	***0.7503***	0.7493	***0.7477***
Range	0.0000	0.0010	***0.0020***	***0.0000***	***0.0030***	0.0020	***0.0010***	***0.0010***	***0.0040***

Group $\overline{X}$ and Range Chart

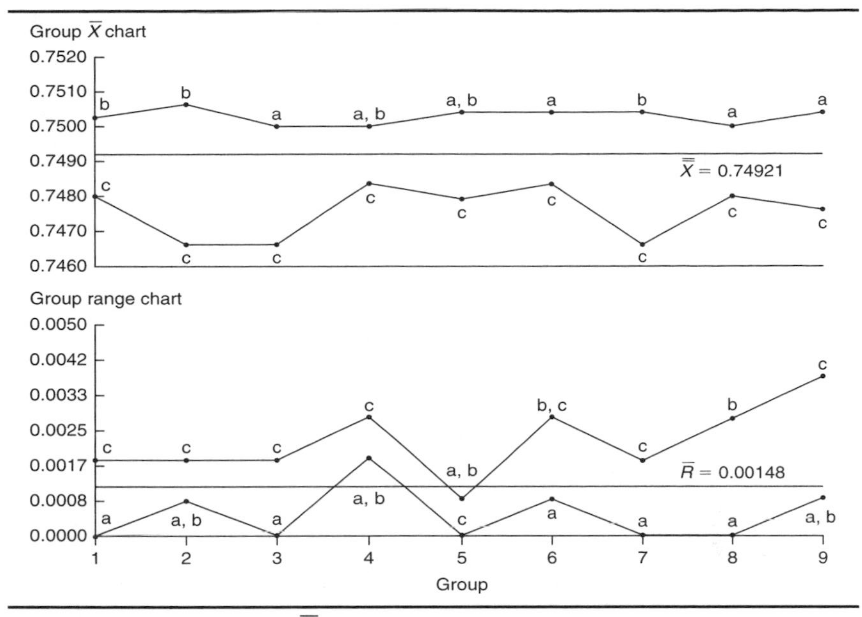

Figure 21.2. Group $\overline{X}$ and range charts representing three ODs on the same part.

Chart Interpretation

Group range chart: Location c appears in the MAX position seven out of nine times. This strongly suggests that location c has the largest standard deviation. Location a appears eight out of nine times in the MIN position, therefore, location a most likely has the smallest standard deviation. The value of location b's standard deviation falls somewhere between the value of the standard deviation of locations a and c.

> **Note:** The centerline on the group range chart is the average of all 27 ranges found in Table 21.5.

Group $\overline{X}$ chart: Locations a and b are in the MAX position six times and five times respectively. This sharing of the MAX position means that the average diameters of a and b behave similarly and they are always larger than location c, which appears nine out of nine times in the MIN position.

> **Note:** The centerline on the group $\overline{X}$ chart is the average of all 27 $\overline{X}$ values found in Table 21.5.

Recommendations

- These charts illustrate the lack of uniformity in the poppet OD. The first recommendation is to change the process so that location c's diameter increases enough to be in line with the size of the diameters at locations a and b. This might be done by reworking the cam or changing the program on the screw machine.
- The large amount of variation at location c should also be addressed. To do this, operators might try a different way of positioning the work piece material in the holding fixture or find a different way to machine the dimension at location c.

Estimating the Process Average

Process average estimates should be performed separately for each characteristic or location on the group chart (see Calculation 21.1).

$$\overline{\overline{X}}_a = \frac{\sum \overline{X}_a}{k_a} = \frac{6.7502}{9} = 0.75002$$

Where k_a is the number of diameter a subgroups in the data set.

Calculation 21.1. Estimate of the process average for location a.

Estimating σ

Estimates of σ are also calculated separately for each characteristic or location on the group chart. Continuing with location a, see Calculations 21.2 and 21.3.

$$\overline{R}_a = \frac{\sum R_a}{k_a} = \frac{0.006}{9} = 0.00067$$

Calculation 21.2. Calculation of $\overline{R}$ for location a.

$$\hat{\sigma}_a = \frac{\overline{R}_a}{d_2} = \frac{0.00067}{1.693} = 0.00039$$

Calculation 21.3. Estimated standard deviation for location a.

Note: To ensure reliable estimates, k needs to be at least 20. In this example, k is only nine. Therefore, these estimates and the ones found in Table 21.6 are shown only for illustration purposes.

Calculating Process Capability and Performance Ratios

Calculations 21.4, 21.5, and 21.6 show the process capability and performance calculations for location a.

$$C_{pa} = \frac{USL_a - LSL_a}{6\hat{\sigma}_a} = \frac{0.7530 - 0.7470}{6(0.00039)} = \frac{0.0060}{0.00234} = 2.56$$

Calculation 21.4. C_p calculation for location a.

$$C_{pk_u a} = \frac{USL_a - \overline{\overline{X}}_a}{3\hat{\sigma}_a} = \frac{0.7530 - 0.75002}{3(0.00039)} = \frac{0.00298}{0.00117} = 2.55$$

Calculation 21.5. $C_{pk\ upper}$ calculation for location a.

$$C_{pk_l a} = \frac{\overline{\overline{X}}_a - LSL_a}{3\hat{\sigma}_a} = \frac{0.75002 - 0.7470}{3(0.00039)} = \frac{0.00302}{0.00117} = 2.58$$

Calculation 21.6. $C_{pk\ lower}$ calculation for location a.

Group $\overline{X}$ and Range Chart Advantages

- Multiple characteristics can be tracked on one chart.
- Pinpoints the characteristics that are in need of the most attention.
- Separates variation due to changes in the average from variation due to changes in the standard deviation.

Group $\overline{X}$ and Range Chart Disadvantages

- No visibility of the characteristics that fall between the MAX and MIN plot points
- Cannot detect certain out-of-control conditions because the group charts described here have no control limits

An Additional Comment About the Case

The process capability and performance calculations for locations b and c are shown in Table 21.6.

Table 21.6. Additional summary statistics and process capability and performance ratios.

Location b	Location c
$\overline{\overline{X}}_b = 0.75000$	$\overline{\overline{X}}_c = 0.74760$
$\overline{R}_b = 0.00156$	$\overline{R}_c = 0.00222$
$\hat{\sigma}_b = 0.00092$	$\sigma_c = 0.00131$
$C_{p\,b} = 1.09$	$C_{p\,c} = 0.76$
$C_{pk_u\,b} = 1.09$	$C_{pk_u\,c} = 1.37$
$C_{pk_l\,b} = 1.09$	$C_{pk_l\,c} = 0.15$

Chapter 22

Group $\overline{X}$ and s Chart

Decision Tree Section

Table 22.1. Control chart decision tree.

Number of characteristics or locations on the same chart?	Subgroup size?	Different averages on the same chart?	Different standard deviations on the same chart?	Use this chart
>1	≥10	No	No	Group $\overline{X}$ and s chart

Description

Group $\overline{X}$ *Chart*

The group $\overline{X}$ chart is used to monitor and detect changes in the averages among multiple measurement locations of an identical characteristic. At each location, several measurements are taken and an average is calculated. Averages for multiple locations are placed into a group. The plot points on the group $\overline{X}$ chart represent the MAX and MIN averages from the group.

For best results, the number of measurement locations in a group should not exceed five. The number of measurements used to calculate an average at each location is typically 10 or more when combining the group $\overline{X}$ chart with the group s chart.

Group s Chart

The group s chart is used to monitor and detect changes in the standard deviation among multiple measurement locations of an identical characteristic. At each location, several measurements are taken and a sample standard deviation (s) value is calculated. Values of s for multiple locations are placed into a group. The plot points on the group s chart represent the MAX and MIN s values from the group.

Subgroup Assumptions

- Independent measurements within a subgroup
- Constant sample size
- One characteristic
- One unit of measure

Calculating Plot Points

Table 22.2. Formulas for calculating plot points on the group $\overline{X}$ and s chart.

Chart	Plot point	Plot point formula
Group $\overline{X}$	MAX and MIN $\overline{X}$	$\frac{\Sigma X}{n}$
Group s	MAX and MIN s	$\sqrt{\frac{\Sigma(x_i - \overline{X})^2}{n-1}}$

Calculating Centerlines

Table 22.3. Formulas for calculating centerlines on the group $\overline{X}$ and s chart.

Chart	Centerline	Centerline formula
Group $\overline{X}$	$\overline{\overline{X}}$	$\frac{\Sigma \overline{X}}{\text{(Number of groups)(Number of characteristics per group)}}$
Group s	$\overline{s}$	$\frac{\Sigma s}{\text{(Number of groups)(Number of characteristics per group)}}$

Calculating Control Limits

Table 22.4. Control limits for the group $\overline{X}$ and *s* chart.

Chart	Upper control limit	Lower control limit
Group $\overline{X}$	none	none
Group *s*	none	none

Example

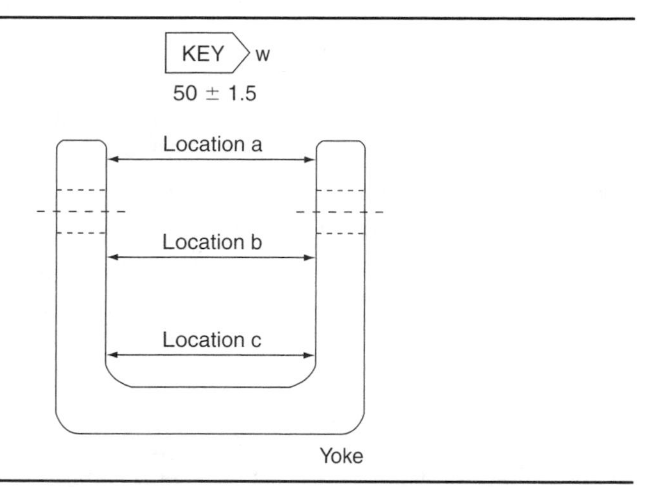

Figure 22.1. Three width measurements from a yoke.

Case Description

This yoke is machined from an aluminum casting. There have been complaints from the assembly department that some of the yokes have a taper on the inside width. To monitor the uniformity of the inside widths, a group chart is set up at the milling machine to track the width at locations a, b, and c.

Sampling Strategy

Because the production volume is very high, and the same characteristic is being measured at three different locations on the part, a group $\overline{X}$ and *s* chart is selected. Ten yokes are measured every hour.

Data Collection Sheet

Table 22.5. Data collection sheet for the group $\overline{X}$ and *s* chart. MAX and MIN plot points are shown in bold.

Group number	1			2			3		
Location	a	b	c	a	b	c	a	b	c
Sample 1	48.8	49.4	49.5	49.0	50.2	49.7	48.4	49.4	49.3
2	49.6	49.7	49.8	50.5	50.2	49.5	50.6	49.1	49.4
3	49.8	49.3	50.2	51.4	49.7	50.1	46.3	50.7	50.3
4	49.5	49.5	49.1	48.1	48.5	49.4	50.0	49.3	50.0
5	50.6	50.0	49.1	50.3	49.5	48.2	49.9	48.6	49.8
6	49.5	49.7	48.9	47.2	49.2	49.3	49.1	49.3	49.0
7	48.8	49.7	49.7	49.6	50.0	49.6	49.5	49.4	49.7
8	46.9	49.7	49.8	49.7	50.3	49.2	51.0	49.2	48.7
9	49.6	50.3	49.3	48.2	49.3	50.5	51.2	49.4	49.9
10	49.8	49.7	49.5	49.1	49.6	49.8	49.5	49.5	49.7
$\overline{X}$	***49.29***	***49.70***	49.49	***49.31***	***49.65***	49.53	49.55	***49.39***	***49.58***
s	***0.98***	***0.29***	0.40	***1.26***	***0.56***	0.61	***1.43***	0.53	***0.48***

Group number	4			5			6		
Location	a	b	c	a	b	c	a	b	c
Sample 1	46.8	50.0	49.1	47.1	50.3	50.2	48.9	49.7	49.3
2	51.3	49.6	49.8	45.9	50.2	49.6	49.8	49.5	49.5
3	51.0	49.2	49.4	49.4	48.4	49.7	48.8	49.6	49.4
4	50.0	49.5	50.1	49.6	50.2	50.0	48.2	49.6	49.9
5	48.5	50.3	49.7	50.0	49.7	49.2	51.0	49.9	49.9
6	50.4	49.7	49.2	51.6	49.9	49.3	49.0	49.4	50.3
7	49.3	49.3	50.2	49.6	49.1	49.3	51.8	49.4	49.8
8	47.6	50.1	48.9	48.3	50.6	49.7	49.7	49.6	49.8
9	49.9	48.8	49.3	51.4	50.1	49.6	47.5	50.8	49.6
10	46.9	48.5	49.5	48.8	48.4	49.6	47.9	50.5	49.9
$\overline{X}$	***49.17***	49.50	***49.52***	***49.17***	***49.69***	49.62	***49.26***	***49.80***	49.74
s	***1.64***	0.57	***0.43***	***1.76***	0.79	***0.31***	***1.35***	0.48	***0.30***

Table 22.5. Cont.

Group number	7			8			9		
Location	a	b	c	a	b	c	a	b	c
Sample 1	49.4	49.3	49.2	50.6	50.0	48.7	51.0	49.0	50.0
2	47.1	49.9	50.4	46.9	49.4	49.3	50.0	48.7	48.2
3	50.5	49.1	49.8	50.4	49.5	49.2	49.2	48.8	48.9
4	50.4	49.1	49.4	50.1	49.8	49.7	49.2	48.5	49.4
5	47.1	49.3	50.1	52.3	50.0	49.2	49.0	49.4	49.7
6	48.0	49.8	49.2	49.5	49.4	49.8	48.5	49.6	48.5
7	51.8	49.4	49.4	47.7	49.4	49.4	49.7	49.1	49.8
8	49.9	49.6	49.0	48.8	48.9	49.5	48.6	50.3	50.1
9	52.5	49.3	49.8	49.6	49.3	50.3	48.4	49.6	49.1
10	50.0	48.4	48.4	50.3	49.4	49.0	48.5	48.8	48.8
$\overline{X}$	***49.67***	***49.32***	49.47	***49.62***	49.51	***49.41***	49.21	***49.18***	***49.25***
s	***1.83***	***0.42***	0.58	***1.54***	***0.34***	0.45	***0.83***	***0.55***	0.65

Group $\overline{X}$ and s Chart

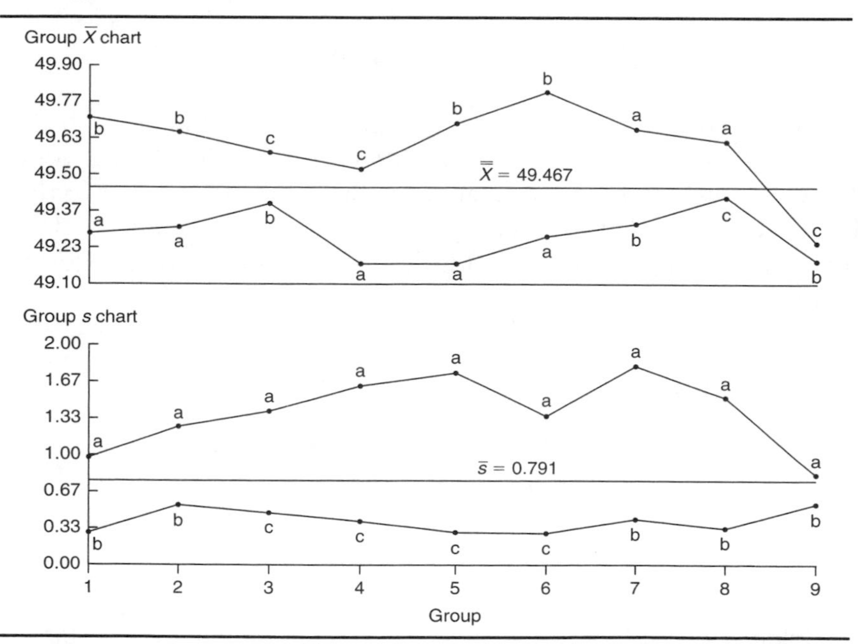

Figure 22.2. Group $\overline{X}$ and *s* chart representing three different yoke width locations.

Chart Interpretation

Group s *chart:* Location a appears in the MAX position for all groups. This suggests that location a has the largest standard deviation. Locations b and c appear randomly in the MIN position. This indicates that locations b and c have similar standard deviations and they are less than location a's.

> **Note:** The centerline on the group *s* chart is the average of all the *s* values on the data collection sheet.

Group $\overline{X}$ *chart:* The difference between the MAX and MIN for each group represents taper within the yokes. Locations a, b, and c appear randomly in the MAX position. However, location a appears five out of nine times in the MIN position. This might indicate that location a has a smaller diameter than either of the two other locations. However, this supposition is not as strong as it would be if location a represented the MIN position for all groups.

> **Note:** The centerline on the group $\overline{X}$ chart is the average of all the $\overline{X}$ plot points found on the data collection sheet.

Recommendation

The repeated presence of location a in the MAX position in the group *s* chart may be the result of the inability of tooling to hold the workpiece consistently during the manufacturing of the yokes. Notice that location a is found at the end of the yoke. This may signify the need for tooling changes that will hold the outer ends more rigidly during manufacturing.

Estimating the Process Average

Process average estimates should be performed separately for each characteristic or location on the group chart (see Calculation 22.1).

$$\overline{\overline{X}}_a = \frac{\Sigma \overline{X}_a}{\text{Number of groups}} = \frac{444.25}{9} = 49.361$$

Calculation 22.1. Estimate of the process average for yoke width at location a.

Estimating σ

Estimates of σ are also calculated separately for each characteristic or location on the group chart. Continuing with yoke width location a, see Calculations 22.2 and 22.3.

$$\bar{s}_a = \frac{\sum s_a}{\text{Number of groups}} = \frac{12.62}{9} = 1.402$$

Calculation 22.2. Calculation of the average sample standard deviation for yoke width location a.

$$\hat{\sigma}_a = \frac{\bar{s}_a}{c_4} = \frac{1.402}{0.9727} = 1.442$$

Calculation 22.3. Estimated standard deviation for yoke width location a.

Note: To ensure reliable estimates, the number of groups should be at least 20. In this example, the number of groups is only nine. Therefore, these estimates and those found in Table 22.6 are only for illustration purposes.

Calculating Process Capability and Performance Ratios

Calculations 22.4, 22.5, and 22.6 show the process capability and performance calculations for yoke width location a.

$$C_{pa} = \frac{\text{USL}_a - \text{LSL}_a}{6\hat{\sigma}_a} = \frac{51.50 - 48.50}{6(1.442)} = \frac{3.00}{8.652} = 0.35$$

Calculation 22.4. C_p calculation for width location a.

$$C_{pk_u a} = \frac{\text{USL}_a - \overline{\overline{X}}_a}{3\hat{\sigma}_a} = \frac{51.50 - 49.361}{3(1.442)} = \frac{2.139}{4.326} = 0.49$$

Calculation 22.5. $C_{pk\ upper}$ calculation for width location a.

$$C_{pk_l a} = \frac{\overline{\overline{X}}_a - \text{LSL}_a}{3\hat{\sigma}_a} = \frac{49.361 - 48.50}{3(1.442)} = \frac{0.861}{4.326} = 0.20$$

Calculation 22.6. $C_{pk\ lower}$ calculation for width location a.

Group $\overline{X}$ *and* s *Chart Advantages*

- Graphically illustrates the variation of multiple product or process characteristics simultaneously and relative to each other.

- Pinpoints the characteristics that are in need of the most attention.
- Separates variation due to changes in the average from variation due to changes in the standard deviation.
- Multiple measurement locations can be tracked on one chart.

Group $\overline{X}$ and s Chart Disadvantages

- No visibility of the characteristics that fall between the MAX and MIN plot points.
- Cannot detect certain out-of-control conditions because the group charts described here have no control limits.
- Given the large amounts of data used in *s* charts, efficient analysis typically requires software.

An Additional Comment About the Case

The process capability and performance ratio calculations for yoke widths at locations b and c are shown in Table 22.6.

Table 22.6. Summary statistics and process capability and performance ratios for yoke widths at locations b and c.

Width location b	Width location c
$\overline{\overline{X}}_b = 49.527$	$\overline{\overline{X}}_c = 49.512$
$\bar{s}_b = 0.503$	$\bar{s}_c = 0.467$
$\hat{\sigma}_b = 0.517$	$\hat{\sigma}_c = 0.481$
$C_{p\,b} = 0.97$	$C_{p\,c} = 1.04$
$C_{pk_u\,b} = 1.27$	$C_{pk_u\,c} = 1.38$
$C_{pk_l\,b} = 0.66$	$C_{pk_l\,c} = 0.70$

Chapter 23

Charts for Multiple Characteristics with Different Targets — The Group Target Charts

In this chapter, three charts will be introduced that will help SPC practitioners manage situations where they encounter the need to monitor multiple related characteristics, each with different targets, on the same chart.

Let's take, for example, a chemical bath where concentrations are being monitored. The chemical bath of interest contains five concentrations to be tested. The target concentration level for each chemical is different (see Table 23.1), and management would like to evaluate all concentrations on one chart.

Table 23.1. Concentration targets for five different chemicals in a single process bath.

Chemical	Target concentration
A	15.0%
B	8.5%
C	5.5%
D	5.5%
E	4.0%

When monitoring multiple characteristics on the same chart, the use of a group chart is suggested. When monitoring characteristics with different means on the same chart, data coding (values defined as deviation from target) is required. For the chemical concentration example just described, one would benefit from using a group target chart.

The three group target charts that will be addressed in the following chapters are

1. Group target individual X and moving range chart
2. Group target $\overline{X}$ and range chart
3. Group target $\overline{X}$ and s chart

Each of these group target charts is similar to their regular target chart counterparts (see chapter 11) in that one is able to evaluate the distance between a measurement and its target. So, the IX and $\overline{X}$ charts have a zero point that represents the target value for each different characteristic.

However, each of the group target charts also has the advantage of evaluating multiple characteristics. Instead of having only one data point line like the regular target charts, the group target charts incorporate two lines. The top line represents the maximum values in each group and the bottom line represents the minimum values. The values falling between the maximum and minimum plot points are not shown for the sake of simplicity and readability.

Chapter 24

Group Target Individual X and Moving Range Chart

Decision Tree Section

Table 24.1. Control chart decision tree.

Number of characteristics or locations on the same chart?	Subgroup size?	Different averages on the same chart?	Different standard deviations on the same chart?	Use this chart
>1	1	Yes	No	Group target individual X and moving range chart

Description

Group Target Individual X *Chart*

The group target individual X chart is used to monitor and detect changes in the process average among multiple measurement locations of similar types of characteristics. The characteristics may come from different part numbers and may have different engineering nominal values, but should all be produced by the same process. The individual measurements from these multiple characteristics and locations are coded by subtracting target values (usually engineering nominal) from individual measurements and combining the coded individual values into a logical group. While the plot points represent the MAX and MIN coded measurements in a group, the sample size, n, is one. For the group target IX chart to be most effective, the number of individual measurements in a group should not exceed five.

Group Moving Range Chart

The group moving range chart is used to monitor and detect changes in the standard deviation among multiple measurement locations of similar types of characteristics. Moving ranges from these multiple locations are combined into a logical group. The plot points represent MAX and MIN moving ranges between consecutive groups.

Subgroup Assumptions

- Independent measurements
- Sample size of one ($n = 1$)
- Similar characteristics
- One unit of measure

Calculating Plot Points

Table 24.2. Formulas for calculating plot points for the group target *IX* and *MR* chart.

Chart	Plot point	Plot point formula
Group target individual *X*	MAX and MIN coded individual *X*	Individual X – Target
Group moving range	MAX and MIN moving range	Moving range between two consecutive measurements from the same location

Calculating Centerlines

Table 24.3. Formulas for calculating centerlines for the group target *IX* and *MR* chart.

Chart	Centerline	Centerline formula
Group target *IX*	Coded $\overline{IX}$	$\dfrac{\Sigma \text{ coded } IX}{(\text{Number of groups})(\text{Number of characteristics per group})}$
Group *MR*	$\overline{MR}$	$\dfrac{\Sigma MR}{(\text{Number of groups} - \text{Number of unique parts}) \times (\text{Number of characteristics per group})}$

Calculating Control Limits

Table 24.4. Control limits for the group target *IX* and *MR* chart.

Chart	Upper control limit	Lower control limit
Group target *IX*	none	none
Group *MR*	none	none

Example

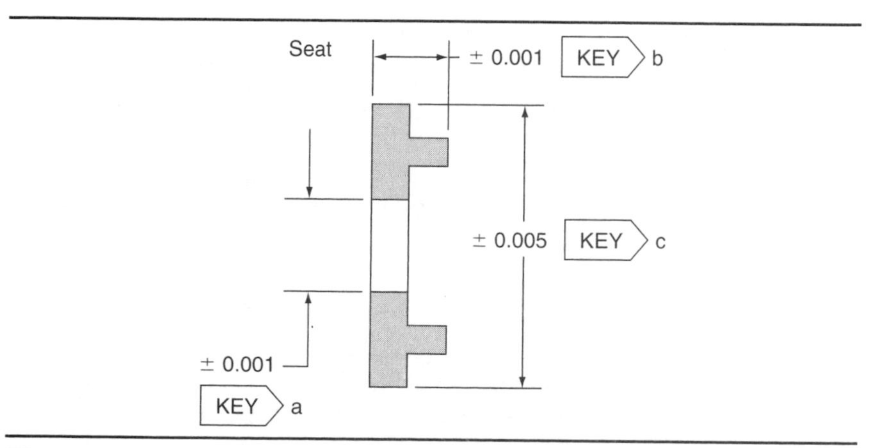

Figure 24.1. Three generic key characteristics for the seat product line.

Case Description

Three generic key characteristics are monitored on several different seat products. All seats share three common key characteristics and tolerances.

- Key a, inside diameter (nominal ± 0.001)
- Key b, length (nominal ± 0.001)
- Key c, OD (nominal ± 0.005)

Seats are manufactured in many different sizes. In this example, three different seat product series (the -400, -800, and -900) will be evaluated. Each of the three seat series is machined on the same lathe but with different tools. Each characteristic is a different size, but the standard deviations are expected to be similar. The shop supervisor wants to analyze the stability of all three key characteristics, regardless of series number, on one chart (see Table 24.5).

Table 24.5. Key target values for the three different seat product series.

Series	Key a target	Key b target	Key c target
-400	0.4450	0.3240	0.8555
-800	0.5000	0.3640	0.9580
-900	0.5700	0.4050	1.0700

Sampling Strategy

Given low production volume and multiple characteristics of different sizes, a group target individual *X* and moving range chart is selected. This chart will help operators evaluate the variation due to the lathe and variation specific to each characteristic/product series combination. The data in Table 24.6 represent measurements taken at the lathe every hour in subgroup sizes of one.

Data Collection Sheet

Table 24.6. Group target *IX* and *MR* data and plot points (shown in bold) for the three seat product line characteristics.

Seat series	-400			-400		
Group number	1			2		
Key characteristic	a	b	c	a	b	c
Individual *X*	0.4448	0.3246	0.8549	0.4455	0.3238	0.8551
Target	0.4450	0.3240	0.8555	0.4450	0.3240	0.8555
Individual *X* − Target	−0.0002	***0.0006***	***−0.0006***	***0.0005***	−0.0002	***−0.0004***
Moving range	—	—	—	0.0007	***0.0008***	***0.0002***

Seat series	-400			-800		
Group number	3			4		
Key characteristic	a	b	c	a	b	c
Individual *X*	0.4451	0.3241	0.8557	0.4997	0.3636	0.9581
Target	0.4450	0.3240	0.8555	0.5000	0.3640	0.9580
Individual *X* − Target	***0.0001***	***0.0001***	***0.0002***	−0.0003	***−0.0004***	***0.0001***
Moving range	0.0004	***0.0003***	***0.0006***	—	—	—

Table 24.6. Cont.

Seat series	-800			-800		
Group number	5			6		
Key characteristic	a	b	c	a	b	c
Individual X	0.4997	0.3637	0.9581	0.4998	0.3640	0.9585
Target	0.5000	0.3640	0.9580	0.5000	0.3640	0.9580
Individual X – Target	**–0.0003**	**–0.0003**	**0.0001**	**–0.0002**	0.0000	**0.0005**
Moving range	**0.0000**	**0.0001**	**0.0000**	**0.0001**	0.0003	**0.0004**
Seat series	-900			-900		
Group number	7			8		
Key characteristic	a	b	c	a	b	c
Individual X	0.5700	0.4050	1.0699	0.5699	0.4049	1.0701
Target	0.5700	0.4050	1.0700	0.5700	0.4050	1.0700
Individual X – Target	**0.0000**	**0.0000**	**–0.0001**	**–0.0001**	**–0.0001**	**0.0001**
Moving range	—	—	—	**0.0001**	**0.0001**	**0.0002**
Seat series	-900					
Group number	9					
Key characteristic	a	b	c			
Individual X	0.5702	0.4051	1.0702			
Target	0.5700	0.4050	1.0700			
Individual X – Target	**0.0002**	**0.0001**	**0.0002**			
Moving range	**0.0003**	0.0002	**0.0001**			

Plot Point Calculation

Group moving range chart: Moving range values are calculated by taking the absolute value between individual measurements from consecutive groups for the same location. For example, location a in group 2 is 0.4455 and location a in group 1 is 0.4448, so the moving range between the two groups is $|0.4455 - 0.4448| = 0.0007$. MAX and MIN values within each group are used as plot points.

Note: Because the same part series was not evaluated in any previous group, no moving range values exist for groups 1, 4, or 7.

Group Target IX *and Moving Range Chart*

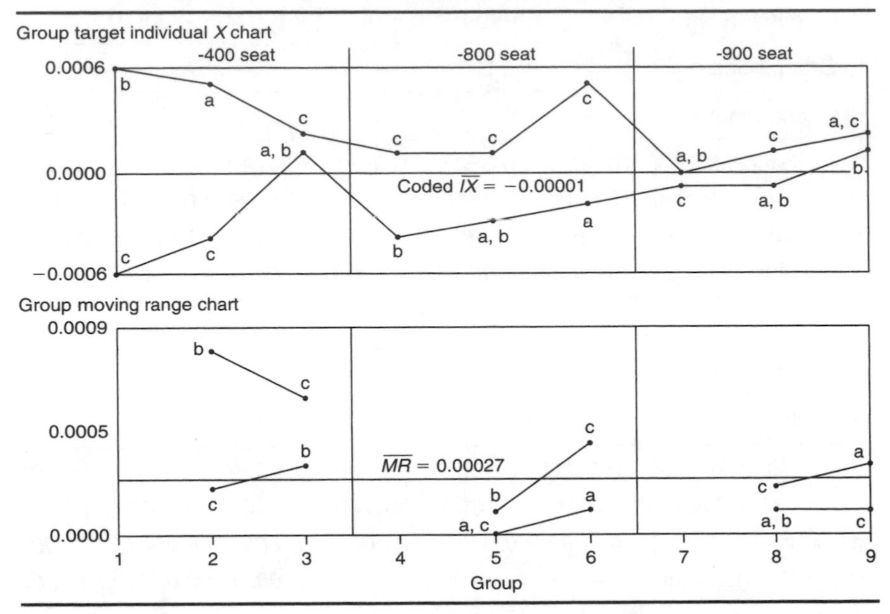

Figure 24.2. Group target *IX* and moving range chart used to evaluate three different key characteristics from three similar parts.

Chart Interpretation

Note: There are only three groups per part series in this example, therefore, any plot point patterns unique to a part series should be considered only when more data become available.

Group moving range chart: MAX and MIN plot points from consecutive groups appear to be descending over time. This could be the result of either

1. The standard deviation getting smaller over time regardless of part number
2. The -400 series parts exhibiting more variability than either the -800 or -900 series seats

With more data, this initial observation could be confirmed or rejected.

Note: The centerline on the group moving range chart is the average of all the moving ranges in the data set.

Group target individual X *chart:* Key characteristic c appears in the MAX position six out of nine times. Because this is true across all three part series, it may indicate a condition inherent to the process instead of one specific to a part series. Operators speculate it has to do with the lathe's apparent difficulty in machining the ODs. There might be something unique about why the lathe tends to run ODs a little higher than specified. Or the problem may be attributed to the programmer having written the program to intentionally manufacture the diameters on the high side. Additional investigation will be required to pinpoint the reason for this nonrandom pattern.

Note: The centerline on the group target *IX* chart is the average of all the coded *IX* plot points in the data set.

Recommendations

- As more data are collected, the operator should pay close attention to key characteristic c (the ODs). Look for reasons why the diameters on all part series might be a little high.
- Look to see if the moving range plot points continue to decrease over time. It is possible that the -400 series key characteristics have larger standard deviations than the -800 or -900 series key characteristics. (The -800 and -900 series are larger parts, which could explain their smaller standard deviations.)

Estimating the Process Average

If all the part series and their characteristics on the *IX* chart appear to be behaving randomly, a single average of all coded individual values could be used to estimate the overall process average. However, because this was not the case for the seat products here, process averages will need to be estimated for each seat characteristic across all part series. This is done by calculating a coded $\overline{IX}$ value for each characteristic for all part series. An example for characteristic a is shown in Calculation 24.1.

$$\text{Coded } \overline{IX}_a = \frac{\sum \text{Coded } IX_a}{\text{Number of groups}}$$

$$= \frac{-0.0003}{9} = -0.000033$$

Calculation 24.1. Estimate of the process average for key characteristic a.

Estimating σ

Estimates of σ are also calculated separately for each characteristic on the group chart. Continuing with key characteristic a, see Calculations 24.2 and 24.3.

$$\overline{MR}_a = \frac{\sum MR_a}{(\text{Number of groups} - \text{Number of parts})}$$
$$= \frac{0.0016}{(9-3)} = \frac{0.0016}{6}$$
$$= 0.00027$$

Calculation 24.2. Calculation of $\overline{MR}$ for key characteristic a across all seat series.

$$\hat{\sigma}_a = \frac{\overline{MR}_a}{d_2} = \frac{0.00027}{1.128} = 0.00024$$

Calculation 24.3. Estimate of the process standard deviation for key characteristic a.

Note: To ensure reliable estimates, the number of groups should be at least 20. In this example, the number of groups is only 9. Therefore, these estimates and those found in Table 24.7 are shown only for illustration purposes.

Calculating Process Capability and Performance Ratios

These ratios are calculated using coded data. The coded target for each characteristic is zero. Calculations for key characteristic a across all three part series are shown in Calculations 24.4, 24.5, and 24.6.

$$C_{pa} = \frac{USL_a - LSL_a}{6\hat{\sigma}_a} = \frac{0.001 - (-0.001)}{6(0.00024)} = \frac{0.002}{0.00144} = 1.39$$

Calculation 24.4. C_p calculation for seat key characteristic a.

$$C_{pk_u a} = \frac{USL_a - \overline{IX}_a}{3\hat{\sigma}_a} = \frac{0.001 - (-0.000033)}{3(0.00024)}$$
$$= \frac{0.001033}{0.00072} = 1.43$$

Calculation 24.5. $C_{pk\ upper}$ calculation for seat key characteristic a.

$$C_{pk_l a} = \frac{\overline{IX}_a - LSL_a}{3\hat{\sigma}_a} = \frac{-0.000033 - (-0.001)}{3(0.00024)}$$

$$= \frac{0.000967}{0.00072} = 1.34$$

Calculation 24.6. $C_{pk\ lower}$ calculation for seat key characteristic a.

Group Target $\overline{\text{IX}}$ and Moving Range Chart Advantages

- Graphically illustrates the variation of multiple products and their characteristics simultaneously on the same chart
- Separates sources of variation unique to the process, unique to the product, and unique to a characteristic on a single chart
- Separates variation due to changes in the average from variation due to changes in the standard deviation

Group Target $\overline{\text{IX}}$ and Moving Range Chart Disadvantages

- No visibility of the characteristics that fall between the MAX and MIN plot points.
- The use of negative numbers can be confusing.
- Cannot detect certain out-of-control conditions because the group charts described here have no control limits.

An Additional Comment About the Case

Additional statistics and process capability and performance values for key characteristics b and c are shown in Table 24.7.

Table 24.7. Additional statistics and process capability and performance values for key characteristics b and c.

Key characteristic b	Key characteristic c
$\overline{IX}_b = -0.000022$	$\overline{IX}_c = 0.000011$
$\overline{MR}_b = 0.0003$	$\overline{MR}_c = 0.00025$
$\hat{\sigma}_b = 0.00027$	$\hat{\sigma}_c = 0.00022$
$C_{p\,b} = 1.25$	$C_{p\,c} = 0.75$
$C_{pk_u\,b} = 1.28$	$C_{pk_u\,c} = 0.74$
$C_{pk_l\,b} = 1.23$	$C_{pk_l\,c} = 0.77$

Chapter 25

Group Target $\overline{X}$ and Range Chart

Decision Tree Section

Table 25.1. Control chart decision tree.

Number of characteristics or locations on the same chart?	Subgroup size?	Different averages on the same chart?	Different standard deviations on the same chart?	Use this chart
>1	>1 <10	Yes	No	Group target $\overline{X}$ and range chart

Description

Group Target $\overline{X}$ *Chart*

The group target $\overline{X}$ chart is used to monitor and detect changes in the process average among multiple measurement locations of similar types of characteristics. The characteristics may come from different part numbers and may have different engineering nominal values, but should all be produced by the same process. Data from these multiple characteristics and locations are coded by subtracting target values (usually engineering nominal) from location averages and combining the coded averages into a logical group. For the group target $\overline{X}$ chart to be most effective, the number of coded averages in a group should not exceed five.

The plot points represent the MAX and MIN coded averages in a grouping. The subgroup size, *n*, used for calculating averages at each location may range from two to nine, but practitioners generally use subgroup sizes of three or five.

Group Range Chart

The group range chart is used to monitor and detect changes in the standard deviation among multiple measurement locations of similar types of characteristics. Ranges from these multiple locations are combined into a logical group. The plot points represent MAX and MIN ranges in the group.

Subgroup Assumptions

- Independent measurements within a subgroup
- Constant sample size
- Similar characteristics
- One unit of measure

Calculating Plot Points

Table 25.2. Formulas used for calculating plot points for the group target $\overline{X}$ and range chart.

Chart	Plot point	Plot point formula
Group target $\overline{X}$	MAX and MIN coded $\overline{X}$	$\left[\frac{\Sigma X}{n}\right] - \text{Target}$
Group range	MAX and MIN range	Largest X − Smallest X in a subgroup

Calculating Centerlines

Table 25.3. Formulas for calculating centerlines on the group target $\overline{X}$ and range chart.

Chart	Centerline	Centerline formula
Group target $\overline{X}$	Coded $\overline{\overline{X}}$	$\dfrac{\sum \text{Coded } \overline{X}}{\text{(Number of groups)(Number of characteristics per group)}}$
Group range	$\overline{R}$	$\dfrac{\sum R}{\text{(Number of groups)(Number of characteristics per group)}}$

Calculating Control Limits

Table 25.4. Control limits on the group target $\overline{X}$ and range chart.

Chart	Upper control limit	Lower control limit
Group target $\overline{X}$	none	none
Group range	none	none

Example

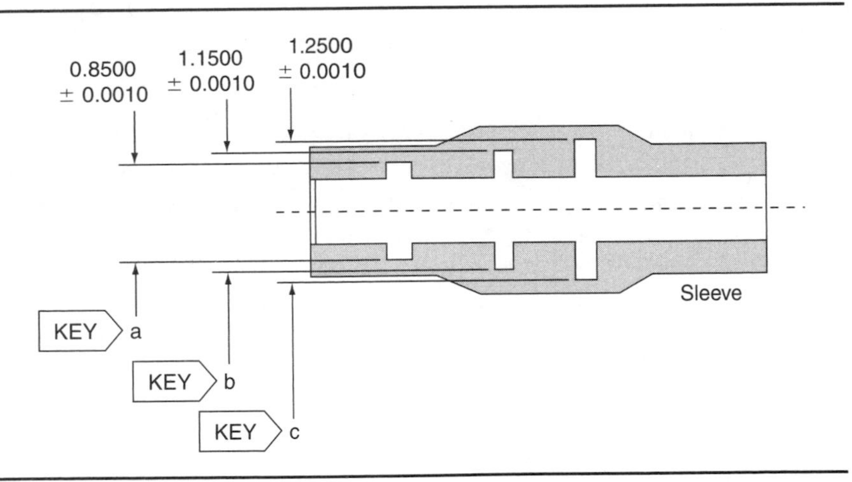

Figure 25.1. Three sleeve inside diameter key characteristics.

Case Description

This sleeve contains three inside diameter key characteristics. They are all machined on the same lathe but with different tools. Each inside diameter is a different size. The customer requires stability of the lathe process as well as capability information from each inside diameter.

Sampling Strategy

Visibility is required of both process and part variability. Because the same type of characteristic (sleeve diameters) with different targets is being measured at multiple locations on the same part, a group target $\overline{X}$ and range chart is selected. This chart will highlight both variation in the lathe and variation in each of the three sleeves.

The cycle time required to manufacture a sleeve is three minutes. Cutting tools are replaced about every two hours. The subgroups represent measurements taken every hour from three consecutive sleeves.

Data Collection Sheet

Table 25.5. Group target $\overline{X}$ and range chart data for three sleeve characteristics. MAX and MIN plot points are shown in bold.

Group number	1			2			3		
Characteristic	a	b	c	a	b	c	a	b	c
Sample 1	0.8500	1.1494	1.2494	0.8507	1.1498	1.2494	0.8509	1.1496	1.2494
2	0.8512	1.1494	1.2489	0.8498	1.1491	1.2488	0.8507	1.1498	1.2490
3	0.8501	1.1491	1.2494	0.8510	1.1499	1.2497	0.8506	1.1497	1.2496
ΣX	2.5513	3.4479	3.7477	2.5515	3.4488	3.7479	2.5522	3.4491	3.7480
$\overline{X}$	0.8504	1.1493	1.2492	0.8505	1.1496	1.2493	0.8507	1.1497	1.2493
Target	0.8500	1.1500	1.2500	0.8500	1.1500	1.2500	0.8500	1.1500	1.2500
$\overline{X}$ − Target	**0.0004**	−0.0007	**−0.0008**	**0.0005**	−0.0004	**−0.0007**	**0.0007**	−0.0003	**−0.0007**
Range	**0.0012**	**0.0003**	0.0005	**0.0012**	**0.0008**	0.0009	0.0003	**0.0002**	**0.0006**

Group number	4			5			6		
Characteristic	a	b	c	a	b	c	a	b	c
Sample 1	0.8504	1.1499	1.2489	0.8512	1.1500	1.2490	0.8511	1.1498	1.2493
2	0.8502	1.1496	1.2493	0.8499	1.1498	1.2496	0.8507	1.1501	1.2497
3	0.8512	1.1497	1.2500	0.8508	1.1497	1.2497	0.8505	1.1498	1.2493

Table 25.5. Cont.

Group number	4			5			6		
Characteristic	a	b	c	a	b	c	a	b	c
ΣX	2.5518	3.4492	3.7482	2.5519	3.4495	3.7483	2.5523	3.4497	3.7483
$\overline{X}$	0.8506	1.1497	1.2494	0.8506	1.1498	1.2494	0.8508	1.1499	1.2494
Target	0.8500	1.1500	1.2500	0.8500	1.1500	1.2500	0.8500	1.1500	1.2500
$\overline{X}$ − Target	***0.0006***	−0.0003	***−0.0006***	***0.0006***	−0.0002	***−0.0006***	***0.0008***	−0.0001	***−0.0006***
Range	0.0010	***0.0003***	***0.0011***	***0.0013***	***0.0003***	0.0007	***0.0006***	***0.0003***	0.0004
Group number	7			8			9		
Characteristic	a	b	c	a	b	c	a	b	c
Sample 1	0.8512	1.1498	1.2490	0.8506	1.1498	1.2495	0.8510	1.1501	1.2495
2	0.8506	1.1494	1.2498	0.8506	1.1501	1.2497	0.8518	1.1498	1.2498
3	0.8505	1.1501	1.2496	0.8512	1.1500	1.2492	0.8499	1.1503	1.2492
ΣX	2.5523	3.4493	3.7484	2.5524	3.4499	3.7484	2.5527	3.4502	3.7485
$\overline{X}$	0.8508	1.1498	1.2495	0.8508	1.1500	1.2495	0.8509	1.1501	1.2495
Target	0.8500	1.1500	1.2500	0.8500	1.1500	1.2500	0.8500	1.1500	1.2500
$\overline{X}$ − Target	***0.0008***	−0.0002	***−0.0005***	***0.0008***	0.0000	***−0.0005***	***0.0009***	0.0001	***−0.0005***
Range	***0.0007***	***0.0007***	***0.0008***	***0.0006***	***0.0003***	0.0005	***0.0019***	***0.0005***	0.0006

Group Target $\overline{X}$ *and Range Chart*

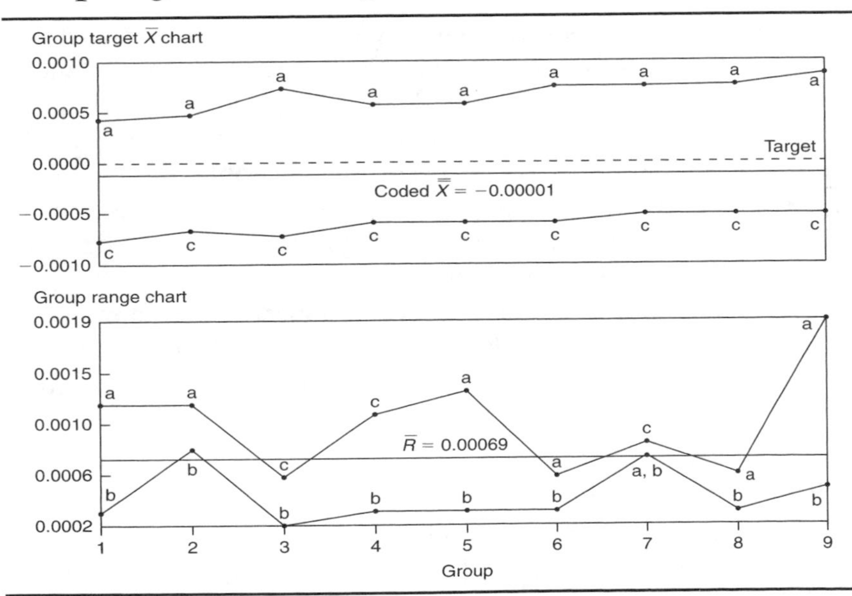

Figure 25.2. Group target $\overline{X}$ and range chart representing three different sized inside sleeve diameters.

Chart Interpretation

Group range chart: Either characteristic a or c shows up in the MAX position in every group. This suggests that these two locations have the largest standard deviation values. Location b appears in the MIN position in every group. This means that, of the three diameters being evaluated, location b has the least variability.

> **Note:** The centerline on the group range chart is the average of all the ranges in the data collection sheet.

Group target $\overline{X}$ chart: Diameter a dominates the MAX position. It consistently deviates from its target (to the high side) more than the other diameters. Location c dominates the MIN position. It consistently deviates from its target (to the low side) more than the other diameters. Diameter b falls in between. It deviates from its target value less than diameters a or c. This is characteristic of taper in the diameters. Also, notice that the MAX and MIN lines are somewhat parallel and seem to gradually trend upwards.

> **Note:** The centerline on the group target $\overline{X}$ chart is the average of all the coded $\overline{X}$ plot points in the data collection sheet.

Recommendations

- Operators should find out why the diameters on the ends (a and c) have larger standard deviations. One might evaluate the cutting tools, the way the sleeve is held when machined, part loading techniques, wall thicknesses at the different locations, coolant flow, or measurement problems.
- People working in the process should try to eliminate the taper among the diameters. Change the process so that the a and c diameters fall closer to their targets.
- The upward trend on the $\overline{X}$ chart appears to be a predictable tool wear condition. One may consider performing a regression analysis to estimate when the cutting tool(s) should be replaced.

Estimating the Process Average

If all of the key characteristics on the group target $\overline{X}$ chart appeared to be behaving randomly, a single estimate of the process average could be used to estimate the process average for all locations. However, in this case, the group target $\overline{X}$ chart does not exhibit random behavior.

Given this nonrandom behavior on the group target $\overline{X}$ chart, estimates of the process average should be calculated separately for each characteristic on the group target chart. This is illustrated in Calculation 25.1 using data from diameter a.

$$\overline{\overline{X}}_a = \frac{\sum \overline{X}_a}{k_a} = \frac{7.6561}{9} = 0.85068$$

Where k_a is the number of diameter a subgroups in the data set.

Calculation 25.1. Estimate of the process average for diameter a.

Estimating σ

Estimates of σ are also calculated separately for each characteristic on the group chart. Continuing with diameter a, see Calculations 25.2 and 25.3.

$$\overline{R}_a = \frac{\sum R_a}{k_a} = \frac{0.0088}{9} = 0.00098$$

Calculation 25.2. Calculation of $\overline{R}$ for use in estimating the process standard deviation for diameter a.

$$\hat{\sigma}_a = \frac{\overline{R}_a}{d_2} = \frac{0.00098}{1.693} = 0.00058$$

Calculation 25.3. The estimate of the process standard deviation for diameter a.

Note: To ensure reliable estimates, *k* needs to be at least 20. In this example, *k* is only nine. Therefore, the estimates here and in Table 25.6 are for illustration purposes only.

Calculating Process Capability and Performance Ratios

Calculations 25.4, 25.5, and 25.6 show the C_p and C_{pk} calculations for diameter a.

$$C_{pa} = \frac{\text{USL}_a - \text{LSL}_a}{6\hat{\sigma}_a} = \frac{0.8510 - 0.8490}{6(0.00058)} = \frac{0.0020}{0.00348} = 0.57$$

Calculation 25.4. C_p calculation for diameter a.

$$C_{pk_u a} = \frac{USL_a - \overline{\overline{X}}_a}{3\hat{\sigma}_a} = \frac{0.8510 - 0.85068}{3(0.00058)} = \frac{0.00032}{0.00174} = 0.18$$

Calculation 25.5. $C_{pk\ upper}$ calculation for diameter a.

$$C_{pk_l a} = \frac{\overline{\overline{X}}_a - LSL_a}{3\hat{\sigma}_a} = \frac{0.85068 - 0.8490}{3(0.00058)} = \frac{0.00168}{0.00174} = 0.97$$

Calculation 25.6. $C_{pk\ lower}$ calculation for diameter a.

Group Target $\overline{X}$ and Range Chart Advantages

- Simultaneously illustrates the variation of multiple product or process characteristics.
- Similar characteristics with different averages can be analyzed on the same chart.
- Separates variation due to changes in the average from variation due to changes in the standard deviation.
- Multiple characteristics can be tracked on one chart.

Group Target $\overline{X}$ and Range Chart Disadvantages

- No visibility of the characteristics that fall between the MAX and MIN plot points.
- The use of negative numbers can be confusing.
- Cannot detect certain nonrandom conditions because the group charts described here have no control limits.

Additional Comments About the Case

- The remaining process statistics and process capability and performance ratios for diameters b and c are shown in Table 25.6.
- Diameter a is not capable. Its average is greater than its target by almost 0.0007".
- Diameter b is capable although its average is more than 0.0002" lower than its target.
- Diameter c is not capable and its average is more than 0.0006" lower than its target.

Table 25.6. Additional statistics and process capability and performance values for diameters b and c.

Diameter b	Diameter c
$\overline{\overline{X}}_b = 1.14977$	$\overline{\overline{X}}_c = 1.24939$
$\overline{R}_b = 0.00041$	$\overline{R}_c = 0.00068$
$\hat{\sigma}_b = 0.00024$	$\hat{\sigma}_c = 0.00040$
$C_{p\,b} = 1.37$	$C_{p\,c} = 0.83$
$C_{pk_u\,b} = 1.69$	$C_{pk_u\,c} = 1.34$
$C_{pk_l\,b} = 1.06$	$C_{pk_l\,c} = 0.32$

Chapter 26

Group Target $\overline{X}$ and s Chart

Decision Tree Section

Table 26.1. Control chart decision tree.

Number of characteristics or locations on the same chart?	Subgroup size?	Different averages on the same chart?	Different standard deviations on the same chart?	Use this chart
>1	≥10	Yes	No	Group target $\overline{X}$ and s chart

Description

Group Target $\overline{X}$ *Chart*

The group target $\overline{X}$ chart is used to monitor and detect changes in the process average among multiple measurement locations of similar types of characteristics. The characteristics may come from different part numbers and may have different engineering nominal values, but should all be produced by the same process. Data from these multiple characteristics and locations are coded by subtracting target values (usually engineering nominal) from location averages and combining the coded averages into a logical group. For the group target $\overline{X}$ chart to be most effective, the number of coded averages in a group should not exceed five.

The plot points represent the MAX and MIN coded averages in a grouping. When combining a group target *s* chart with a group target $\overline{X}$ chart, the number of measurements used to calculate an average at each location is typically 10 or more.

Group s Chart

The group *s* chart is used to monitor and detect changes in the standard deviation among multiple measurement locations of some characteristic. At each location, several measurements are taken and a sample standard deviation (*s*) value is calculated. Values of *s* for multiple locations are placed into a group. The plot points on the group *s* chart represent the MAX and MIN *s* values from within each group.

Subgroup Assumptions

- Independent measurements within a subgroup
- Constant sample size
- Similar characteristics
- One unit of measure

Calculating Plot Points

Table 26.2. Formulas for calculating plot points for the group target $\overline{X}$ and *s* chart.

Chart	Plot point	Plot point formula
Group target $\overline{X}$	MAX and MIN coded $\overline{X}$	$\left[\frac{\Sigma X}{n}\right] - \text{Target}$
Group *s*	MAX and MIN *s*	$\sqrt{\frac{\Sigma(x_i - \overline{X})^2}{n-1}}$

Calculating Centerlines

Table 26.3. Formulas for calculating centerlines on the group target $\overline{X}$ and s chart.

Chart	Centerline	Centerline formula
Group target $\overline{X}$	Coded $\overline{\overline{X}}$	$\dfrac{\sum \text{Coded } \overline{X}}{\text{(Number of groups)(Number of characteristics per group)}}$
Group s	$\bar{s}$	$\dfrac{\sum s}{\text{(Number of groups)(Number of characteristics per group)}}$

Calculating Control Limits

Table 26.4. Control limits for the group target $\overline{X}$ and s chart.

Chart	Upper control limit	Lower control limit
Group target $\overline{X}$	none	none
Group s	none	none

Example

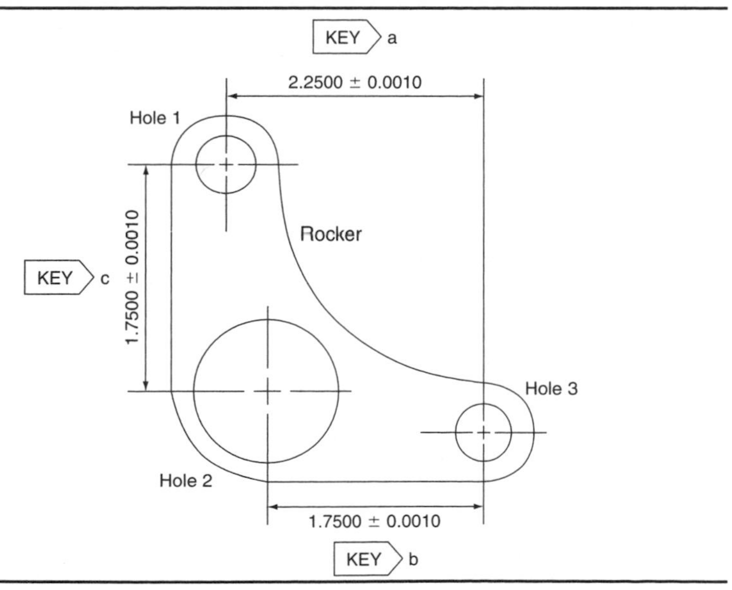

Figure 26.1. Three hole location measurements from a rocker.

Case Description

The rocker shown in Figure 26.1 is machined from an iron casting. There have been complaints from field mechanics that the rockers are not interchangeable and that the holes do not always line up with mating parts. To monitor the uniformity of the hole locations, the operators would like to use a chart at the milling machine to track the variability of the three hole locations.

Sampling Strategy

Because production volume is very high and all the measurements represent hole locations of different distances created on the same machine, a group target $\overline{X}$ and s chart is selected. Ten rockers are measured every hour.

Data Collection Sheet

Table 26.5. Group target $\overline{X}$ and s chart data collection sheet for three hole locations on a rocker. MAX and MIN plot points are shown in bold.

Group number	1			2			3		
Location	a	b	c	a	b	c	a	b	c
Sample 1	2.2499	1.7501	1.7490	2.2499	1.7501	1.7496	2.2506	1.7499	1.7493
2	2.2503	1.7502	1.7494	2.2514	1.7502	1.7495	2.2500	1.7500	1.7488
3	2.2499	1.7500	1.7490	2.2496	1.7499	1.7488	2.2503	1.7502	1.7494
4	2.2500	1.7501	1.7490	2.2503	1.7499	1.7494	2.2508	1.7498	1.7492
5	2.2501	1.7496	1.7493	2.2507	1.7501	1.7495	2.2500	1.7504	1.7494
6	2.2506	1.7500	1.7495	2.2511	1.7501	1.7494	2.2506	1.7497	1.7493
7	2.2507	1.7499	1.7490	2.2505	1.7500	1.7494	2.2507	1.7497	1.7492
8	2.2504	1.7495	1.7495	2.2506	1.7503	1.7491	2.2505	1.7503	1.7486
9	2.2499	1.7498	1.7492	2.2499	1.7497	1.7490	2.2499	1.7498	1.7490
10	2.2496	1.7499	1.7487	2.2504	1.7500	1.7487	2.2502	1.7503	1.7493
$\overline{X}$	2.2501	1.7499	1.7492	2.2504	1.7500	1.7492	2.2504	1.7500	1.7492
Target	2.2500	1.7500	1.7500	2.2500	1.7500	1.7500	2.2500	1.7500	1.7500
$\overline{X}$ − Target	***0.0001***	−0.0001	−***0.0008***	***0.0004***	0.0000	−***0.0008***	***0.0004***	0.0000	−***0.0008***
s	***0.0004***	***0.0002***	0.0003	***0.0006***	***0.0002***	0.0003	***0.0003***	***0.0003***	***0.0003***
Group number	4			5			6		
Location	a	b	c	a	b	c	a	b	c
Sample 1	2.2502	1.7499	1.7490	2.2494	1.7501	1.7492	2.2500	1.7498	1.7493
2	2.2501	1.7501	1.7492	2.2494	1.7499	1.7493	2.2507	1.7499	1.7494
3	2.2499	1.7497	1.7492	2.2506	1.7499	1.7494	2.2506	1.7496	1.7494
4	2.2502	1.7502	1.7492	2.2504	1.7496	1.7493	2.2506	1.7497	1.7492

Table 26.5. Cont.

Group number	4			5			6		
Location	a	b	c	a	b	c	a	b	c
5	2.2502	1.7501	1.7492	2.2505	1.7502	1.7494	2.2510	1.7502	1.7490
6	2.2505	1.7501	1.7489	2.2499	1.7499	1.7491	2.2501	1.7497	1.7492
7	2.2510	1.7496	1.7486	2.2499	1.7504	1.7493	2.2498	1.7501	1.7491
8	2.2510	1.7500	1.7495	2.2496	1.7501	1.7491	2.2503	1.7496	1.7496
9	2.2504	1.7503	1.7494	2.2506	1.7505	1.7492	2.2500	1.7501	1.7493
10	2.2505	1.7502	1.7495	2.2504	1.7500	1.7493	2.2506	1.7501	1.7489
$\overline{X}$	2.2504	1.7500	1.7492	2.2501	1.7501	1.7493	2.2504	1.7499	1.7492
Target	2.2500	1.7500	1.7500	2.2500	1.7500	1.7500	2.2500	1.7500	1.7500
$\overline{X}$ − Target	***0.0004***	0.0000	−***0.0008***	***0.0001***	***0.0001***	−***0.0007***	***0.0004***	−0.0001	−***0.0008***
s	***0.0004***	***0.0002***	0.0003	***0.0005***	0.0003	***0.0001***	***0.0004***	***0.0002***	***0.0002***

Group number	7			8			9		
Location	a	b	c	a	b	c	a	b	c
Sample 1	2.2500	1.7500	1.7491	2.2503	1.7500	1.7489	2.2508	1.7498	1.7490
2	2.2495	1.7501	1.7491	2.2499	1.7497	1.7493	2.2504	1.7495	1.7494
3	2.2503	1.7497	1.7493	2.2501	1.7500	1.7490	2.2498	1.7496	1.7493
4	2.2498	1.7500	1.7493	2.2509	1.7500	1.7492	2.2499	1.7500	1.7493
5	2.2510	1.7502	1.7493	2.2501	1.7500	1.7496	2.2508	1.7505	1.7487
6	2.2498	1.7503	1.7493	2.2504	1.7501	1.7491	2.2504	1.7499	1.7493
7	2.2511	1.7500	1.7495	2.2500	1.7504	1.7492	2.2506	1.7503	1.7492
8	2.2501	1.7495	1.7491	2.2502	1.7495	1.7489	2.2499	1.7502	1.7496
9	2.2503	1.7504	1.7495	2.2506	1.7501	1.7493	2.2513	1.7499	1.7491
10	2.2497	1.7497	1.7491	2.2508	1.7501	1.7493	2.2505	1.7505	1.7494
$\overline{X}$	2.2502	1.7500	1.7493	2.2503	1.7500	1.7492	2.2504	1.7500	1.7492
Target	2.2500	1.7500	1.7500	2.2500	1.7500	1.7500	2.2500	1.7500	1.7500
$\overline{X}$ − Target	***0.0002***	0.0000	−***0.0007***	***0.0003***	0.0000	−***0.0008***	***0.0004***	0.0000	−***0.0008***
s	***0.0005***	0.0003	***0.0002***	***0.0003***	***0.0002***	***0.0002***	***0.0005***	0.0003	***0.0002***

Group Target $\overline{X}$ and s Chart

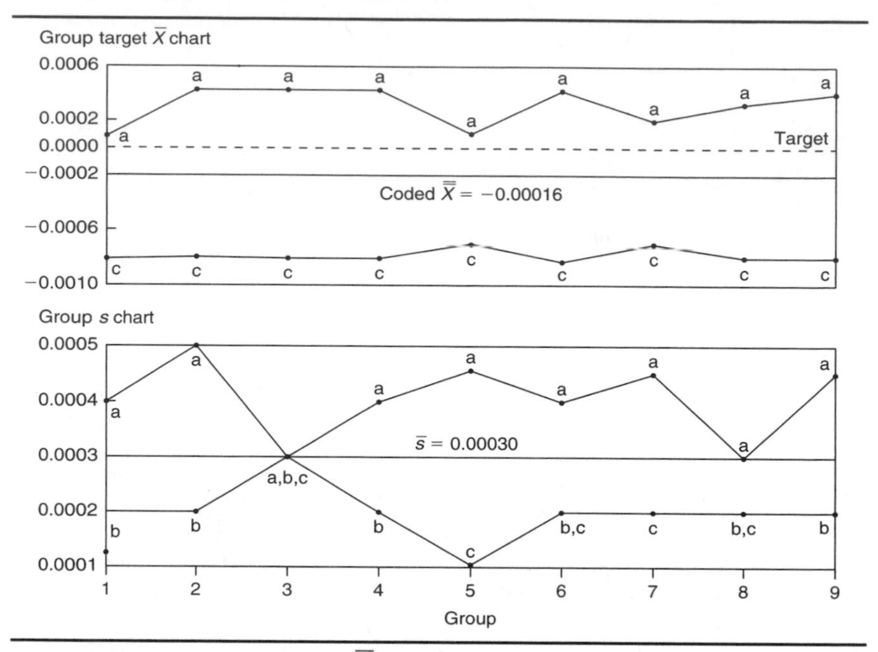

Figure 26.2. Group target $\overline{X}$ and *s* chart representing three different hole locations on the same part.

Chart Interpretation

Group s *chart:* Location a appears in the MAX position in every group. This indicates that location a has the largest standard deviation. Locations b and c appear randomly in the MIN position meaning that location b and c's standard deviation values are both similar to one another and smaller than location a's.

> **Note:** The centerline on the group *s* chart is the average of all the sample standard deviation values on the data collection sheet.

Group target $\overline{X}$ *chart:* The coded $\overline{X}$ for location a appears in the MAX position in every group and its value is always positive. This indicates that the average hole location at location a is consistently higher than the engineering nominal (target) value.

Location c appears in the MIN position in all nine groups and its value is always negative. This means that the average hole location distance at location c is consistently lower than its engineering nominal (target) value.

Note: The centerline on the group target $\overline{X}$ chart is the average of all the coded $\overline{X}$ plot points in the data collection sheet.

Recommendations

- The group target $\overline{X}$ chart reveals two consistent problems: Location a is always wider than target, and location c is always closer. This type of problem is fixed by changing the location of one or more holes during the job setup. The chart itself does not indicate which hole to relocate. A logical place to begin investigation is with hole 1 because its location affects both key locations a and c.
- Looking at the group *s* chart, the distance between holes 1 and 3 (hole location a) varies more than the other hole relationships. This also means there is excess variation in the horizontal axis. Operators should verify this assumption with process engineers and remedy the problem.

Estimating the Process Average

If all of the locations on the group target $\overline{X}$ chart were behaving randomly, a single estimate of the process average could be used to estimate the process average for all locations. However in this case, the group target $\overline{X}$ chart does not exhibit random behavior.

Given nonrandom patterns on a group target $\overline{X}$ chart, estimates of the process average should be calculated separately for each characteristic or location. This is illustrated in Calculation 26.1 using data from hole location a.

$$\overline{\overline{X}}_a = \frac{\sum \overline{X}_a}{\text{Number of groups}} = \frac{20.2527}{9} = 2.25030$$

Calculation 26.1. Estimate of the process average for hole location a.

Estimating σ

Estimates of σ are also calculated separately for each characteristic or location on the group target chart. Continuing with hole location a, see Calculations 26.2 and 26.3.

$$\bar{s}_a = \frac{\sum s_a}{\text{Number of groups}} = \frac{0.0039}{9} = 0.00043$$

Calculation 26.2. Calculation of $\bar{s}$ for use in estimating the process standard deviation for hole location a.

$$\hat{\sigma}_a = \frac{\bar{s}_a}{c_4} = \frac{0.00043}{0.9727} = 0.00044$$

Calculation 26.3. Estimate of the process standard deviation for hole location a.

Note: To ensure reliable estimates, the number of groups should be at least 20. In this example, the number of groups is only nine. Therefore, the estimates here and in Table 26.6 are for illustration purposes only.

Calculating Process Capability and Performance Ratios

The C_p and C_{pk} calculations for hole location a are shown in Calculations 26.4, 26.5, and 26.6.

$$C_{pa} = \frac{\text{USL}_a - \text{LSL}_a}{6\hat{\sigma}_a} = \frac{2.2510 - 2.2490}{6(0.00044)} = \frac{0.002}{0.00264} = 0.76$$

Calculation 26.4. C_p calculation for hole location a.

$$C_{pk_u a} = \frac{\text{USL}_a - \overline{\overline{X}}_a}{3\hat{\sigma}_a} = \frac{2.2510 - 2.25030}{3(0.00044)} = \frac{0.0007}{0.00132} = 0.53$$

Calculation 26.5. $C_{pk\ upper}$ calculation for hole location a.

$$C_{pk_l a} = \frac{\overline{\overline{X}}_a - \text{LSL}_a}{3\hat{\sigma}_a} = \frac{2.25030 - 2.2490}{3(0.00044)} = \frac{0.0013}{0.00132} = 0.98$$

Calculation 26.6. $C_{pk\ lower}$ calculation for hole location a.

Group Target $\overline{X}$ *and* s *Chart Advantages*

- Simultaneously illustrates the variation of multiple product or process characteristics.
- Similar characteristics with different averages can be analyzed on the same chart.

- Separates variation due to changes in the average from variation due to changes in the standard deviation.
- Multiple characteristics can be tracked on one chart.

Group Target $\overline{X}$ *and* s *Chart Disadvantages*

- No visibility of the characteristics that fall between the MAX and MIN plot points.
- The use of negative numbers can be confusing.
- Cannot detect certain non-random conditions because the group target charts described here have no control limits.

An Additional Comment About the Case

The process capability and performance values for hole locations b and c are shown in Table 26.6.

Table 26.6. Summary statistics and process capability and performance ratios for hole locations b and c.

Hole location b	Hole location c
$\overline{\overline{X}}_b = 1.75000$	$\overline{\overline{X}}_c = 1.74922$
$\overline{s}_b = 0.00024$	$\overline{s}_c = 0.00023$
$\hat{\sigma}_b = 0.00025$	$\hat{\sigma}_c = 0.00024$
$C_{p\,b} = 1.33$	$C_{p\,c} = 1.39$
$C_{pk_u\,b} = 1.33$	$C_{pk_u\,c} = 2.47$
$C_{pk_l\,b} = 1.33$	$C_{pk_l\,c} = 0.31$

Chapter 27

Charts for Short Runs and Multiple Characteristics— The Group Short Run Charts

In this section, three more charts will be introduced. These special charts will help the SPC practitioner manage situations where he or she encounters

- Multiple characteristics
- Dissimilar characteristics and standard deviations
- Short production runs
- Limited data collection opportunities

When SPC practitioners are faced with the situations outlined here, they should benefit by using one or a combination of the following charts.

- Group short run individual X and moving range chart
- Group short run $\overline{X}$ and range™ chart
- Group short run $\overline{X}$ and s chart

These charts are useful when one wishes to evaluate multiple process or product characteristics on the same chart when the averages, standard deviations, and even the units of measurement are different.

The group short run charts take into consideration each product's target value, average, and standard deviation and then codes the data into a unitless ratio. This unitless ratio is what allows characteristics of vastly different types to be tracked on the same chart.

Also, because the group short run charts are an adaptation of the short run charts introduced in chapter 15, they have the added advantage of being able to handle processes that produce limited quantities of products and limited opportunities for data collection.

Not only do group short run charts share attributes with short run charts found in chapters 15 through 18, they also share valuable features found in group charts (introduced in chapter 19). That is, for each subgroup of measurements, the MAX and MIN values are identified and plotted on the group short run chart. The MAX values are connected with a line (as are the MIN values) to allow the chart user to visualize within-process variability for individual subgroups.

The benefits of data coding and the use of MAX and MIN values makes the group short run charts especially useful for SPC practitioners who work with complex products and processes that are produced in JIT and single-piece flow manufacturing environments.

Chapter 28

Group Short Run Individual *X* and Moving Range Chart

Decision Tree Section

Table 28.1. Control chart decision tree.

Number of characteristics or locations on the same chart?	Subgroup size?	Different averages on the same chart?	Different standard deviations on the same chart?	Use this chart
>1	1	Yes	Yes	Group short run *IX–MR* chart

Description

The Group Short Run Individual X *Chart*

The group short run individual *X* chart, also called a *group short run IX chart,* is used to monitor and detect changes in individual measurements among multiple characteristics of any type. The characteristics may have different engineering nominal values, different units of measure, and different standard deviations, but should all be related enough to want to analyze them all on the same chart. The plot points represent the MAX and MIN coded individual measurements from a logical grouping. The plot points are coded by subtracting from each measurement a target $\overline{IX}$ (usually the engineering nominal value) and then dividing by a target $\overline{MR}$. Each characteristic on the chart may have a unique target $\overline{IX}$.

The Group Short Run Moving Range Chart

The group short run moving range chart, also called a *group short run MR chart,* is used to monitor and detect changes in the standard deviation among multiple measurement locations and characteristics of any type. Moving ranges are calculated by taking the absolute difference between two consecutive coded *IX* measurements. The plot points represent MAX and MIN moving ranges between groups.

Subgroup Assumptions

- Independent measurements
- Constant sample size

Calculating Plot Points

Table 28.2. Formulas for calculating plot points for the group short run *IX* and *MR* chart.

Chart	Plot points	Plot point formula
Group short run individual *X*	MAX and MIN coded *IX*	$\frac{IX - \text{Target } \overline{IX}}{\text{Target } \overline{MR}}$
Group short run moving range	MAX and MIN coded *MR*	Absolute difference between two consecutive coded *IX* values from the same location

Calculating Centerlines

Table 28.3. Centerlines for the group short run *IX* and *MR* chart.

Chart	Centerline	Centerline formula
Group short run individual *X*	0	0
Group short run moving range	1	1

Calculating Control Limits

Table 28.4. Control limits for the group short run *IX* and *MR* chart.

Chart	Upper control limit	Lower control limit
Group individual *X*	none	none
Group moving range	none	none

Example

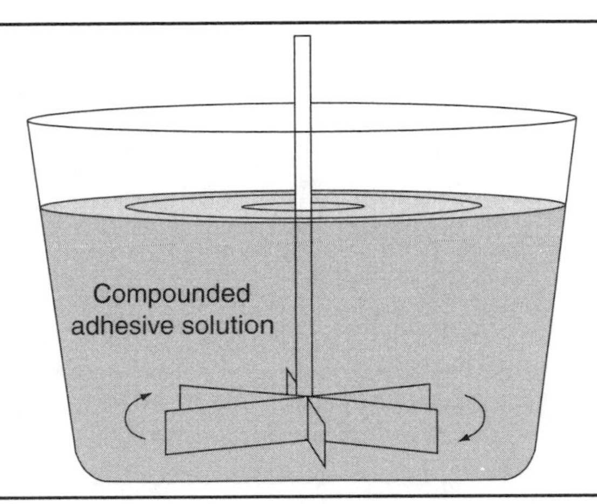

Figure 28.1. Several parameters are monitored for each batch of compounded adhesive solution.

Case Description

The same mixing equipment is used to mix several different types of adhesive compounds. Each compound has its own unique set of functional test requirements. In this example, three compounds are examined: compound A, B, and C. The test requirements for each are listed in Table 28.5.

Table 28.5. Test requirements for compounds A, B, and C.

Compound A				
Characteristic	Chemical 1	Chemical 2	Viscosity	Set time
Specifications	14.5 to 15.5	1.0 to 2.0	11 to 15	15 to 20
Unit of measure	ml/l	%	Seconds	Minutes

Table 28.5. Cont.

Compound B				
Characteristic	Checmical 1	Chemical 2	Hardness	
Specifications	4.5 to 5.0	3.7 to 3.8	72 to 76	
Unit of measure	%	%	Proprietary	
Compound C				
Characteristic	Chemical 1	% solids	Reactant temp	Clarity
Specifications	6.5 to 8.5	14.0 to 15.6	85 to 89	35 to 39
Unit of measure	oz/g	%	°F	Proprietary

Sampling Strategy

The test characteristics, specifications, and units of measure are different for each compound, and only one measurement of each characteristic is gathered from each batch. Therefore, a group short run individual X and moving range chart is selected for use. Target values are established for each characteristic from each compound. The target $\overline{IX}$ values are set at the engineering nominal, but the target $\overline{MR}$ values were derived from quality assurance records.

Target Values

Table 28.6. Target values for compounds A, B, and C.

Compound A				
Characteristic	Chemical 1	Chemical 2	Viscosity	Set time
Target $\overline{IX}$	15	1.50	13	17.5
Target $\overline{MR}$	0.34	0.68	0.46	1.38
Compound B				
Characteristic	Chemical 1	Chemical 2	Hardness	
Target $\overline{IX}$	4.75	3.75	74	
Target $\overline{MR}$	0.13	0.03	0.57	

Table 28.6. Cont.

Compound C				
Characteristic	Chemical 1	% solids	Reactant temp	Clarity
Target $\overline{IX}$	7.5	14.8	86.2	37
Target $\overline{MR}$	0.68	0.23	0.80	2.05

Data Collection Sheet

Table 28.7. Group short run *IX–MR* chart data collection sheet. MAX and MIN plot points are shown in bold.

Compound	A				A			
Group number	1				2			
Parameter	1	2	V	t	1	2	V	t
IX	15.70	1.80	12.90	16.50	15.70	2.00	13.50	16.40
Target $\overline{IX}$	15.00	1.50	13.00	17.50	15.00	1.50	13.00	17.50
IX – Target $\overline{IX}$	0.70	0.30	−0.10	−1.00	0.70	0.50	0.50	−1.10
Target $\overline{MR}$	0.34	0.68	0.46	1.38	0.34	0.68	0.46	1.38
(*IX* – Target $\overline{IX}$)/Target $\overline{MR}$	***2.06***	0.44	−0.22	***−0.72***	***2.06***	0.74	1.09	***−0.80***
Coded *MR*	—	—	—	—	***0.00***	0.30	***1.31***	0.08

Compound	A				A			
Group number	3				4			
Parameter	1	2	V	t	1	2	V	t
IX	15.60	1.70	12.90	18.10	15.30	1.40	13.40	17.20
Target $\overline{IX}$	15.00	1.50	13.00	17.50	15.00	1.50	13.00	17.50
IX – Target $\overline{IX}$	0.60	0.20	−0.10	0.60	0.30	−0.10	0.40	−0.30
Target $\overline{MR}$	0.34	0.68	0.46	1.38	0.34	0.68	0.46	1.38
(*IX* – Target $\overline{IX}$)/Target $\overline{MR}$	***1.76***	0.29	***−0.22***	0.43	***0.88***	−0.15	0.87	***−0.22***
Coded *MR*	***0.30***	0.45	***1.31***	1.23	0.88	***0.44***	***1.09***	0.65

Compound	A				B		
Group number	5				6		
Parameter	1	2	V	t	1	2	H
IX	15.40	2.00	13.60	15.30	4.71	3.80	72.10
Target $\overline{IX}$	15.00	1.50	13.00	17.50	4.75	3.75	74.00
IX – Target $\overline{IX}$	0.40	0.50	0.60	−2.20	−0.04	0.05	−1.90
Target $\overline{MR}$	0.34	0.68	0.46	1.38	0.13	0.03	0.57

Table 28.7. Cont.

Compound	A				B		
Group number	5				6		
Parameter	1	2	V	t	1	2	H
$(IX - \text{Target } \overline{IX})/\text{Target } \overline{MR}$	1.18	0.74	**1.30**	**−1.59**	−0.31	**1.67**	**−3.33**
Coded MR	**0.30**	0.89	0.43	**1.37**	—	—	—

Compound	B			B		
Group number	7			8		
Parameter	1	2	H	1	2	H
IX	4.70	3.78	72.54	4.72	3.73	73.53
Target $\overline{IX}$	4.75	3.75	74.00	4.75	3.75	74.00
$IX - \text{Target } \overline{IX}$	−0.05	0.03	−1.46	−0.03	−0.02	−0.47
Target $\overline{MR}$	0.13	0.03	0.57	0.13	0.03	0.57
$(IX - \text{Target } \overline{IX})/\text{Target } \overline{MR}$	−0.38	***1.00***	***−2.56***	***−0.23***	−0.67	***−0.82***
Coded MR	***0.07***	0.67	***0.77***	***0.15***	1.67	***1.74***

Compound	B			B		
Group number	9			10		
Parameter	1	2	H	1	2	H
IX	4.75	3.74	73.23	4.81	3.74	72.74
Target $\overline{IX}$	4.75	3.75	74.00	4.75	3.75	74.00
$IX - \text{Target } \overline{IX}$	0.00	−0.01	−0.77	0.06	−0.01	−1.26
Target $\overline{MR}$	0.13	0.03	0.57	0.13	0.03	0.57
$(IX - \text{Target } \overline{IX})/\text{Target } \overline{MR}$	***0.00***	−0.33	***−1.35***	***0.46***	−0.33	***−2.21***
Coded MR	***0.23***	0.34	***0.53***	0.46	***0.00***	***0.86***

Compound	C				C			
Group number	11				12			
Parameter	1	S	T	C	1	S	T	C
IX	7.82	14.83	85.23	38.12	7.69	14.66	84.85	37.65
Target $\overline{IX}$	7.50	14.80	86.20	37.00	7.50	14.80	86.20	37.00
$IX - \text{Target } \overline{IX}$	0.32	0.03	−0.97	1.12	0.19	−0.14	−1.35	0.65
Target $\overline{MR}$	0.68	0.23	0.80	2.05	0.68	0.23	0.80	2.05
$(IX - \text{Target } \overline{IX})/\text{Target } \overline{MR}$	0.47	0.13	***−1.21***	***0.55***	0.28	−0.61	***−1.69***	***0.32***
Coded MR	—	—	—	—	***0.19***	***0.74***	0.48	0.23

Table 28.7. Cont.

Compound	C				C			
Group number	13				14			
Parameter	1	S	T	C	1	S	T	C
IX	7.94	14.89	85.80	36.82	7.34	15.11	84.60	35.87
Target $\overline{IX}$	7.50	14.80	86.20	37.00	7.50	14.80	86.20	37.00
IX – Target $\overline{IX}$	0.44	0.09	−0.40	−0.18	−0.16	0.31	−1.60	−1.13
Target $\overline{MR}$	0.68	0.23	0.80	2.05	0.68	0.23	0.80	2.05
(*IX* – Target $\overline{IX}$)/Target $\overline{MR}$	***0.65***	0.39	***−0.50***	−0.09	−0.24	***1.35***	***−2.00***	−0.55
Coded *MR*	***0.37***	1.00	***1.19***	0.41	0.89	0.96	***1.50***	***0.46***

Compound	C				C			
Group number	15				16			
Parameter	1	S	T	C	1	S	T	C
IX	8.07	14.65	84.59	38.01	7.02	14.66	84.39	39.66
Target $\overline{IX}$	7.50	14.80	86.20	37.00	7.50	14.80	86.20	37.00
IX – Target $\overline{IX}$	0.57	−0.15	−1.61	1.01	−0.48	−0.14	−1.81	2.66
Target $\overline{MR}$	0.68	0.23	0.80	2.05	0.68	0.23	0.80	2.05
(*IX* – Target $\overline{IX}$)/Target $\overline{MR}$	***0.84***	−0.65	***−2.01***	0.49	−0.71	−0.61	***−2.26***	***1.30***
Coded *MR*	1.08	***2.00***	***0.01***	1.04	***1.55***	***0.04***	0.25	0.81

Group Short Run Individual X *and Moving Range Chart*

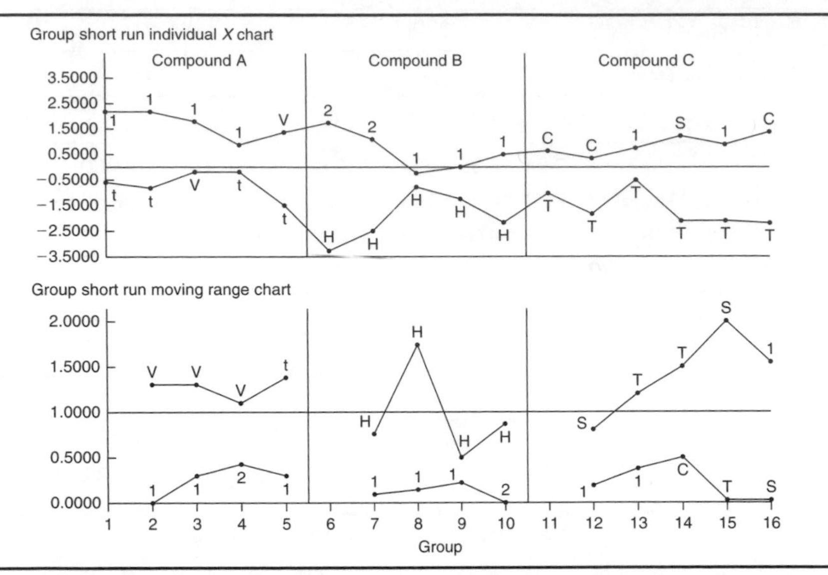

Figure 28.2. Group short run *IX* and *MR* chart for three different compounded adhesive solutions.

Chart Interpretation

Group short run moving range chart: There is a run of four consecutive hardness (*H*) plot points in the MAX position from compound B. This indicates that there is significantly more variation in the hardness characteristic than others.

Also, the first 10 MIN plot points are all chemical concentrations. This indicates that the chemical concentration characteristics exhibit the lowest variability of the characteristics being evaluated regardless of the compound.

Group short run individual X *chart:* All but one of the MAX plot points from compounds A and B represent chemical concentrations. This means that the chemical concentrations are higher on average than their targets.

All of the MIN plot points for compound B represent the hardness characteristic (*H*). This run indicates that the average hardness is less than its target.

The set time (*t*) from compound A is in the MIN position four out of five times. This may indicate that the set time is generally quicker than its target time of 17.5 minutes.

Lastly, the reactant temperature from compound C is consistently in the MIN position indicating lower than target temperatures.

Recommendations

- The MIN plot point run of chemical concentrations on the moving range chart appears to be significant. It indicates that the standard deviations are consistently less than expected by the established target $\overline{MR}$. Therefore, identify the cause for this improvement and update the target values.
- The MAX plot point run of chemical concentrations on the individual X chart appears to be significant. The actual concentrations are consistently higher than expected by the target $\overline{IX}$. Therefore, identify the cause(s) for these high concentrations and bring them closer to target. However, if the concentration levels were intentionally run high, the target $\overline{IX}$ values should be updated to reflect the desired concentration levels.
- The hardness (H) of compound B found on the group short run IX chart is consistently less than its target $\overline{IX}$. Therefore, identify the cause and change the process to bring the hardness closer to target.
- The set time for compound A is a little faster than its target value. If this is an improvement, update the target.
- The reactant temperature (T) of compound C on the group short run IX chart is consistently less than its target value. Process personnel should attempt to do what is necessary to bring the temperature back up to target or determine if the present temperature level is desirable. If it is desirable, then the target temperature value should be updated.

Estimating the Process Average

Estimates of the process average are calculated separately for each characteristic of each compound on the short run group chart. This is illustrated in Calculation 28.1 using the percent solids (S) from compound C.

$$\overline{IX}_S = \frac{\sum IX_S}{k_S}$$

$$= \frac{14.83 + 14.66 + 14.89 + 15.11 + 14.65 + 14.66}{6}$$

$$= \frac{88.80}{6} = 14.800$$

Calculation 28.1. Estimate of the process average percent solids content(s) from compound C.

Estimating σ

In estimating σ, calculations must be performed separately for each characteristic of each compound on the group short run chart. Notice, however, that no moving ranges have been calculated—only coded *MR* values are shown in Table 28.7.

Moving range values should be calculated using consecutive *IX* values just as is done with traditional *IX* and *MR* charts. So, in Calculations 28.2 and 28.3 and in Table 28.8, standard *MR* values have been used in calculating estimates of σ.

$$\overline{MR}_S = \frac{\sum MR_S}{k_S - 1} = \frac{0.17 + 0.23 + 0.22 + 0.46 + 0.01}{6 - 1}$$

$$= \frac{1.09}{5} = 0.218$$

Calculation 28.2. Average moving range for percent solids content from compound C to be used in the estimate of process standard deviation in Calculation 28.3.

$$\hat{\sigma}_S = \frac{\overline{MR}_S}{d_2} = \frac{0.218}{1.128} = 0.193$$

Calculation 28.3. Estimate of the process standard deviation for percent solids content from compound C.

Note: To ensure reliable estimates of both the process average and process standard deviation, *k* needs to be at least 20. In this example, *k* is only six. Therefore, the estimates here and in Table 28.8 are shown only for illustration purposes.

Calculating Process Capability and Performance Ratios

$$C_{ps} = \frac{\text{USL}_s - \text{LSL}_s}{6\hat{\sigma}_s} = \frac{15.6 - 14.0}{6(0.193)} = \frac{1.6}{1.158} = 1.38$$

Calculation 28.4. C_p for percent solids content from compound C.

$$C_{pk_{us}} = \frac{\text{USL}_s - \overline{IX}_s}{3\hat{\sigma}_s} = \frac{15.6 - 14.800}{3(0.193)} = \frac{0.800}{0.579} = 1.38$$

Calculation 28.5. C_{pku} for percent solids content from compound C.

$$C_{pk_{l}s} = \frac{\overline{IX}_s - \text{LSL}_s}{3\hat{\sigma}_s} = \frac{14.8 - 14.0}{3(0.193)} = \frac{0.80}{0.579} = 1.38$$

Calculation 28.6. C_{pkl} for percent solids content from compound C.

Group Short Run IX–MR *Chart Advantages*

- Graphically illustrates the variation of multiple products and their characteristics simultaneously on the same chart.
- Characteristics from different parts with different means, different standard deviations, and different units of measure can all be analyzed on the same chart.
- Illustrates variation of the process and variation of the specific products.

Group Short Run IX–MR *Chart Disadvantages*

- No visibility of the characteristics that fall between the MAX and MIN plot points.
- $\overline{IX}$, $\overline{MR}$, and estimates of σ must be calculated separately for each characteristic on the chart.
- Analysis and recommendations can be tricky if target origins are not known.

Additional Comments About the Case

- Additional statistics and process capability and performance calculations for compound C's chemical 1, clarity, and reactant temperature are shown in Table 28.8.
- The largest cause for compound C's rejections is due to reactant temperature failures. Based on the $C_{pk\,l}$ of -0.06, more than 50 percent of the batches will fall below the lower specification. With failures this large, one of two actions ought to be considered.
  1. Change the process to ensure a higher average reactant temperature. This might be a good time to perform a designed experiment to help determine what to change in the process.
  2. Reexamine the need for the LSL to remain at 85°. If it can be lowered without compromising adhesive performance, change the specification and allow the average temperature to remain at its current level of 84.91°.

Table 28.8. Process capability and performance calculations for compound C's chemical 1, clarity, and reactant temperature.

Chemical 1	Clarity	Temperature
$\overline{IX}_1 = 7.647$	$\overline{IX}_c = 37.688$	$\overline{IX}_T = 84.910$
$\overline{MR}_1 = 0.552$	$\overline{MR}_c = 1.208$	$\overline{MR}_T = 0.548$
$\hat{\sigma}_1 = 0.489$	$\hat{\sigma}_c = 1.071$	$\hat{\sigma}_T = 0.486$
$C_{p\,1} = 0.68$	$C_{p\,c} = 0.62$	$C_{p\,T} = 1.37$
$C_{pk_u\,1} = 0.58$	$C_{pk_u\,c} = 0.41$	$C_{pk_u\,T} = 2.81$
$C_{pk_l\,1} = 0.78$	$C_{pk_l\,c} = 0.84$	$C_{pk_l\,T} = -0.06$

Chapter 29

Group Short Run $\overline{X}$ and Range™ Chart

Decision Tree Section

Table 29.1. Control chart selection tree.

Number of characteristics or locations on the same chart?	Subgroup size?	Different averages on the same chart?	Different standard deviations on the same chart?	Use this chart
>1	>1 <10	Yes	Yes	Group short run $\overline{X}$ and range™ chart

Description

Group Short Run $\overline{\text{X}}$ *Chart*

The group short run $\overline{X}$ chart is used to monitor and detect changes in the process average among multiple characteristics of any type. The characteristics may have different nominals, different units of measure, and different standard deviations, but should all be related enough to warrant analyzing them on the same chart. The plot points represent coded MAX and MIN averages from a logical grouping of data. Plot points are coded by subtracting from each subgroup average its respective target $\overline{\overline{X}}$ (usually the engineering nominal value) and then dividing by its target $\overline{R}$.

For the group chart to be most effective, there should be no more than five measurement locations within a group. Subgroup sizes for locations within a group may range from two to nine but practitioners generally use subgroup sizes of three or five.

Group Short Run Range Chart

The group short run range chart is used to monitor and detect changes in the standard deviation among multiple characteristics of any type. Range values are first calculated by subtracting the smallest measurement value within a subgroup from its largest measurement value. Ranges from multiple locations are coded by dividing each subgroup range by its respective target $\overline{R}$. The plot points represent coded MAX and MIN range values within a group.

Subgroup Assumptions

- Independent measurements within a subgroup
- Constant sample size

Calculating Plot Points

Table 29.2. Plot point formulas for the group short run $\overline{X}$ and range™ chart.

Chart	Plot point	Plot point formula
Group short run $\overline{X}$	MAX and MIN coded $\overline{X}$	$\frac{\overline{X} - \text{Target } \overline{\overline{X}}}{\text{Target } \overline{R}}$
Group short range	MAX and MIN coded range	$\frac{R}{\text{Target } \overline{R}}$

Calculating Centerlines

Table 29.3. Centerlines for the group short run $\overline{X}$ and range™ chart.

Chart	Centerline	Centerline formula
Group short run $\overline{X}$	0	0
Group short run range	1	1

Calculating Control Limits

Table 29.4. Control limits for the group short run $\overline{X}$ and range™ chart.

Chart	Upper control limit	Lower control limit
Group short run $\overline{X}$	none	none
Group short run range	none	none

Example

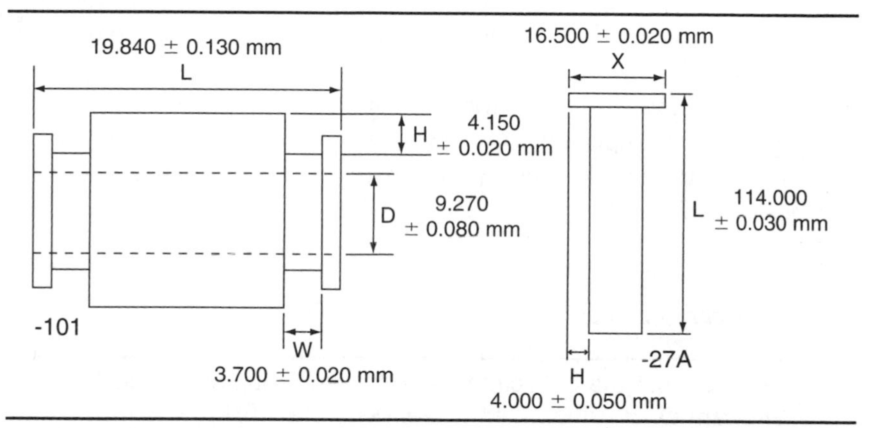

Figure 29.1. Two parts containing multiple key characteristics.

Table 29.5. Key characteristics with respective target values.

Part number	Key characteristic	Target $\overline{\overline{X}}$	Target $\overline{R}$
-101	H	4.150	0.005
	L	19.840	0.038
	D	9.270	0.022
	W	3.700	0.005
-27A	L	114.000	0.038
	H	4.000	0.014
	X	16.500	0.005

Case Description

A single lathe produces many different part numbers, each with many different key characteristics. The two parts shown in Figure 29.1 are examples. The manager of the machine shop wants to use only one chart for each lathe to monitor the process regardless of the part numbers or key characteristics being produced.

Sampling Strategy

The same chart must allow for different part numbers and different key characteristics. Because each characteristic may be unique with respect to its nominal, tolerance, and unit of measure, a group short run $\overline{X}$ and range™ chart is selected. This chart will separate variation due to the lathe from variation unique to each part and characteristic.

The cycle time varies, but lot sizes are typically 20 to 100 parts. Cutting tools are replaced about every three hours. The data represent measurements taken every fifteenth part regardless of the part number (n = 3).

Data Collection Sheet

Table 29.6. Data collection sheet for the group short run $\overline{X}$ and range™ chart lathe example. MAX and MIN plot points are shown in bold.

Part number	-101				-101			
Group number	1				2			
Characteristic	H	L	D	W	H	L	D	W
1	4.151	19.865	9.230	3.706	4.151	19.863	9.235	3.713
2	4.147	19.871	9.259	3.704	4.150	19.890	9.246	3.697
3	4.154	19.845	9.233	3.715	4.149	19.860	9.262	3.694
$\overline{X}$	4.151	19.860	9.241	3.708	4.150	19.871	9.248	3.701
Target $\overline{\overline{X}}$	4.150	19.840	9.270	3.700	4.150	19.840	9.270	3.700
$\overline{X}$ − Target $\overline{\overline{X}}$	0.001	0.020	−0.029	0.008	0.000	0.031	−0.022	0.001
Target $\overline{R}$	0.005	0.038	0.022	0.005	0.005	0.038	0.022	0.005
$\frac{\overline{X} - \text{Target } \overline{\overline{X}}}{\text{Target } \overline{R}}$	0.133	0.535	***−1.333***	***1.667***	0.000	***0.816***	***−1.015***	0.267
R	0.007	0.026	0.029	0.011	0.002	0.030	0.027	0.019
R/Target $\overline{R}$	1.400	***0.684***	1.318	***2.200***	***0.400***	0.789	1.227	***3.800***

Table 29.6. Cont.

Part number	-101				-27A		
Group number	3				4		
Characteristic	H	L	D	W	L	H	X
1	4.149	19.866	9.266	3.705	114.051	4.002	16.501
2	4.150	19.886	9.252	3.685	114.059	3.993	16.498
3	4.152	19.867	9.254	3.697	114.039	4.004	16.508
$\bar{X}$	4.150	19.873	9.257	3.696	114.050	4.000	16.502
Target $\bar{\bar{X}}$	4.150	19.840	9.270	3.700	114.000	4.000	16.500
$\bar{X}$ − Target $\bar{\bar{X}}$	0.000	0.033	−0.013	−0.004	0.050	0.000	0.002
Target $\bar{R}$	0.005	0.038	0.022	0.005	0.038	0.014	0.005
$\frac{\bar{X} - \text{Target } \bar{\bar{X}}}{\text{Target } \bar{R}}$	0.067	***0.868***	−0.576	***−0.867***	***1.307***	***−0.024***	0.467
R	0.003	0.020	0.014	0.020	0.020	0.011	0.010
R/Target $\bar{R}$	0.600	***0.526***	0.636	***4.000***	***0.526***	0.786	***2.000***

Part number	-27A			-27A		
Group number	5			6		
Characteristic	L	H	X	L	H	X
1	114.046	3.993	16.499	114.055	4.004	16.503
2	114.024	3.993	16.497	114.086	4.022	16.498
3	114.062	4.002	16.497	114.077	4.013	16.499
$\bar{X}$	114.044	3.996	16.498	114.073	4.013	16.500
Target $\bar{\bar{X}}$	114.000	4.000	16.500	114.000	4.000	16.500
$\bar{X}$ − Target $\bar{\bar{X}}$	0.044	−0.004	−0.002	0.073	0.013	0.000
Target $\bar{R}$	0.038	0.014	0.005	0.038	0.014	0.005
$\frac{\bar{X} - \text{Target } \bar{\bar{X}}}{\text{Target } \bar{R}}$	***1.158***	−0.286	***−0.467***	***1.912***	0.929	***0.000***
R	0.038	0.009	0.002	0.031	0.018	0.005
R/Target $\bar{R}$	***1.000***	0.643	***0.400***	***0.816***	***1.286***	1.000

Part number	-27A		
Group number	7		
Characteristic	L	H	X
1	114.035	4.016	16.498
2	114.056	4.010	16.499
3	114.054	4.003	16.492

Table 29.6. Cont.

Part number	-27A		
Group number	7		
Characteristic	L	H	X
$\overline{X}$	114.048	4.010	16.496
Target $\overline{\overline{X}}$	114.000	4.000	16.500
$\overline{X}$ – Target $\overline{\overline{X}}$	0.048	0.010	–0.004
Target $\overline{R}$	0.038	0.014	0.005
$\frac{\overline{X} - \text{Target } \overline{\overline{X}}}{\text{Target } \overline{R}}$	***1.272***	0.690	***–0.733***
R	0.021	0.013	0.007
R/Target $\overline{R}$	***0.553***	0.929	***1.400***

Group Short Run $\overline{X}$ and Range™ Chart

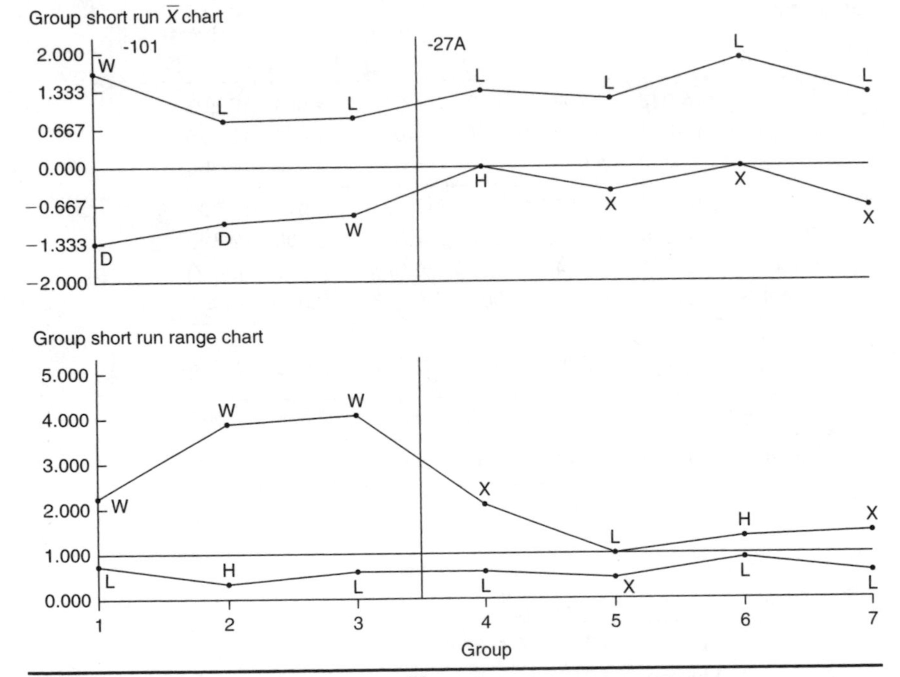

Figure 29.2. Group short run $\overline{X}$ and range™ charts representing two parts and multiple characteristics.

Chart Interpretation

Group short run range chart: During the -101 part run, key characteristic *W* appears in the MAX position all three times. There is a possibility of this happening by chance if all four keys are behaving randomly about their target values, but this may be an indicator of significantly greater variability in the *W* dimension as compared with others.

The *L* dimension appears in the MIN position five out of seven times. This likely represents a nonrandom pattern indicating less variability in the *L* dimension across both parts.

Group short run $\overline{X}$ chart: The length (*L*) characteristic on both the -101 and -27A appears in the MAX position six out of seven times. The chance of this occurring randomly is very small. This is most likely a non-random pattern that is related to the process itself. That is, regardless of the part number, the process tends to cut lengths on the high side.

During the manufacture of the -27A part, the run of three plot points in the MIN position for dimension *X* may indicate the presence of a nonrandom pattern.

Recommendations

- Operators and process engineers should try to identify why the lathe tends to cut all part lengths on the high side and why the *W* dimension on the -101 part displays more relative variation than the other three key characteristics. In addition, operators and engineers should try to isolate the reason why the *L* dimension varies less than other dimensions.
- Watch the *X* dimension on the -27A part and subsequent part numbers. If the dimension continues to fall in the MIN position on the group short run $\overline{X}$ chart, there should be an investigation for nonrandom patterns that relate to process-specific causes. If, however, the *X* dimension fails to fall into the MIN position for subsequent part numbers, the cause should be treated as product specific.

Estimating the Process Average

Estimates of the process average are calculated separately for each characteristic for each part on the group short run charts. This is illustrated in Calculation 29.1 using data from the H dimension on the -27A part.

$$\overline{\overline{X}}_H = \frac{\sum \overline{X}_H}{k_H} = \frac{16.0190}{4} = 4.0048$$

Where k_H is the number of height (H) subgroups in the data set

Calculation 29.1. Estimate of the process average for characteristic *H* on part -27A.

Estimating σ

Estimates of σ are also calculated separately for each characteristic on each part on the group short run charts. Continuing with characteristic H, see Calculations 29.2 and 29.3.

$$\overline{R}_H = \frac{\sum R_H}{k_H} = \frac{0.051}{4} = 0.0128$$

Calculation 29.2. $\overline{R}$ calculation for characteristic *H* on part -27A.

$$\hat{\sigma}_H = \frac{\overline{R}_H}{d_2} = \frac{0.0128}{1.693} = 0.0076$$

Calculation 29.3. Estimate of the process standard deviation for characteristic *H* on part -27A.

Note: To ensure reliable estimates, k needs to be at least 20. In this example, k is only four. Therefore, the estimates shown here and in Table 29.7 are used only for illustration purposes.

Calculating Process Capability and Performance Ratios

Calculations 29.4, 29.5, and 29.6 show the process capability and performance calculations for characteristic H.

$$C_{pH} = \frac{\text{USL}_H - \text{LSL}_H}{6\hat{\sigma}_H} = \frac{4.050 - 3.950}{6(0.0076)} = \frac{0.100}{0.0456} = 2.19$$

Calculation 29.4. C_p calculation for characteristic *H*.

$$C_{pk_u H} = \frac{USL_H - \overline{\overline{X}}_H}{3\hat{\sigma}_H} = \frac{4.050 - 4.0048}{3(0.0076)} = \frac{0.0452}{0.0228} = 1.98$$

Calculation 29.5. $C_{pk\ upper}$ calculation for characteristic *H*.

$$C_{pk_l H} = \frac{\overline{\overline{X}}_H - LSL_H}{3\hat{\sigma}_H} = \frac{4.0048 - 3.950}{3(0.0076)} = \frac{0.0548}{0.0228} = 2.40$$

Calculation 29.6. $C_{pk\ lower}$ calculation for characteristic *H*.

Group Short Run $\overline{X}$ *and Range™ Chart Advantages*

- Graphically illustrates the variation of multiple product or process characteristics on the same chart.
- Characteristics from different parts with different means, different standard deviations, and different units of measure can be analyzed all on the same chart.
- Separates variation due to changes in the average from variation due to changes in the standard deviation.
- Separates variation due to the process from variation specific to a product characteristic.

Group Short Run $\overline{X}$ *and Range™ Chart Disadvantages*

- No visibility of characteristics that fall between the MAX and MIN plot points.
- Cannot detect certain out-of-control conditions because the group charts described here have no control limits.
- Many calculations are required to code the data.

Additional Comments About the Case

- Additional statistics and process capability and performance calculations for part characteristic *L* and *X* for part -27A are shown in Table 29.7.
- Notice that characteristic *L*, while not capable, has a negative $C_{pku\ L}$ value. This indicates that $\overline{\overline{X}}_L$ falls outside of the upper specification limit. In fact, the average falls more than 0.020 mm outside of the USL of 114.03 mm. This underscores the importance of reacting to characteristic *L*'s nonrandom pattern shown on the group short run $\overline{X}$ chart in Figure 29.2.

- Characteristic X has C_p and C_{pk} values that are not only greater than one, but very close, numerically to one another. Therefore, characteristic X is capable and its $\overline{\overline{X}}_X$ is almost perfectly centered on its enginering nominal value of 16.500 mm.

Table 29.7. Additional statistics and process capability and performance ratios for characteristics L and X from part -27A.

Characteristic L	Characteristic X
$\overline{\overline{X}}_L = 114.0538$	$\overline{\overline{X}}_X = 16.4990$
$\overline{R}_L = 0.0275$	$\overline{R}_X = 0.0060$
$\hat{\sigma}_L = 0.0162$	$\hat{\sigma}_X = 0.0035$
$C_{p\,L} = 0.62$	$C_{p\,X} = 1.88$
$C_{pk_u\,L} = -0.49$	$C_{pk_u\,X} = 1.98$
$C_{pk_l\,L} = 1.72$	$C_{pk_l\,X} = 1.79$

Chapter 30

Group Short Run $\overline{X}$ and s Chart

Decision Tree Section

Table 30.1. Control chart decision tree.

Number of characteristics or locations on the same chart?	Subgroup size?	Different averages on the same chart?	Different standard deviations on the same chart?	Use this chart
>1	≥10	Yes	Yes	Group short run $\overline{X}$ and s chart

Description

Group Short Run $\overline{X}$ *Chart*

The group short run $\overline{X}$ chart is used to monitor and detect changes in the process average among multiple characteristics of any type. The characteristics may have different engineering nominals, different units of measure, and different standard deviations, but should all be related enough to warrant analyzing them on the same chart. The plot points represent coded MAX and MIN averages from a logical grouping of data. Plot points are coded by subtracting from each subgroup average its respective target $\overline{\overline{X}}$ (usually the engineering nominal value) and then dividing by its target $\bar{s}$.

For the group chart to be most effective, there should be no more than five measurement locations within a group. When using the group short run $\overline{X}$ chart in conjunction with a group short run s chart, subgroup sizes for locations within a group are typically 10 or more.

Group Short Run s *Chart*

The group short run s chart is used to monitor and detect changes in the standard deviation among multiple characteristics of any type. Sample standard deviation (s) values are first calculated for some product characteristic and then combined into a logical group. After being combined into a group, the s values are coded by dividing each sample standard deviation by its corresponding target $\bar{s}$. The plot points represent MAX and MIN coded s values from a grouping.

Subgroup Assumptions

- Independent measurements within a subgroup
- Constant sample size

Calculating Plot Points

Table 30.2. Plot point formulas for the group short run $\overline{X}$ and s chart.

Chart	Plot point	Plot point formula
Group short run $\overline{X}$	MAX and MIN coded $\overline{X}$	$\frac{\overline{X} - \text{Target } \overline{\overline{X}}}{\text{Target } \bar{s}}$
Group short run s	MAX and MIN coded s	$\frac{s}{\text{Target } \bar{s}}$

Calculating Centerlines

Table 30.3. Centerlines for the group short run $\overline{X}$ and s chart.

Chart	Centerline	Centerline formula
Group short run $\overline{X}$	0	0
Group short run s	1	1

Calculating Control Limits

Table 30.4. Control limits for the group short run $\overline{X}$ and s chart.

Chart	Upper control limit	Lower control limit
Group short run $\overline{X}$	none	none
Group short run s	none	none

Example

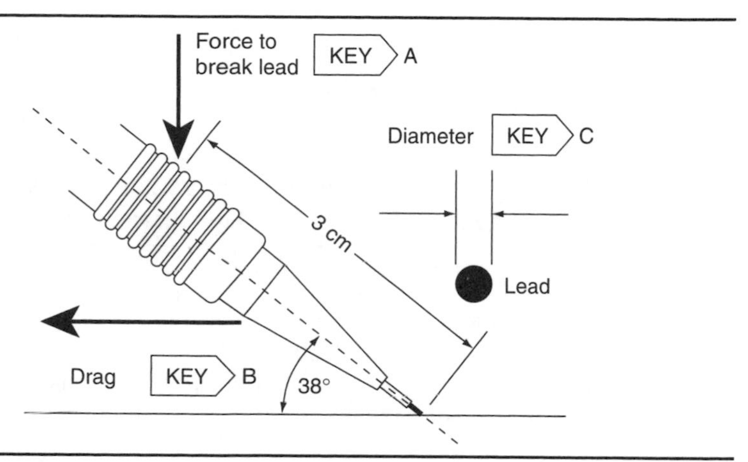

Figure 30.1. Mechanical pencil with three key characteristics.

Case Description

A company manufactures mechanical pencil lead. There are three key characteristics (see Table 30.5).

1. *Break force*—The amount of pressure it takes to break the lead (extended 1.5 mm) at a 38° angle with the force applied 3 cm from the lead tip
2. *Drag*—A proprietary measure of how smoothly the lead releases onto a given paper
3. *Diameter*—The actual diameter of the lead itself

Table 30.5. Upper and lower specification limits for three mechanical pencil lead key characteristics.

Key characteristic	USL	LSL
A (Break force)	2.80 lbs.	1.20 lbs.
B (Drag)	3.510 points	2.490 points
C (Diameter)	0.730 mm	0.670 mm

The manager wishes to monitor the stability of all three key characteristics on the same chart.

Sampling Strategy

Because production volume is very high and three different characteristics are to be monitored, a group short run $\overline{X}$ and s chart is selected. Ten leads are tested every 30 minutes.

Target Values

Preliminary tests on all three key characteristics were conducted. The purpose of the tests was to establish target values for the group short run charts to be used. The target values are found in Table 30.6.

Table 30.6. Target $\overline{\overline{X}}$ and target $\overline{s}$ values for the three mechanical pencil lead key characteristics.

Key characteristic	Target $\overline{\overline{X}}$	Target $\overline{s}$
A (Break force)	2.125 lbs.	0.234 lbs.
B (Drag)	3.000 points	0.080 points
C (Diameter)	0.705 mm	0.005 mm

Data Collection Sheet

Table 30.7. Data collection sheet for the group short run $\overline{X}$ and s chart pencil lead example. MAX and MIN plot points are shown in bold.

Group number	1			2			3		
Characteristic	A	B	C	A	B	C	A	B	C
1	2.004	3.030	0.704	2.040	2.976	0.711	1.924	2.937	0.704
2	2.152	2.995	0.707	2.302	2.986	0.712	2.319	3.044	0.701
3	2.179	3.094	0.703	2.464	3.123	0.706	1.561	2.984	0.716
4	2.130	2.981	0.705	1.879	3.073	0.696	2.208	3.090	0.703
5	2.326	2.806	0.710	2.264	3.048	0.705	2.204	3.035	0.697
6	2.129	2.968	0.707	1.720	2.923	0.702	2.057	2.952	0.703
7	2.009	3.010	0.707	2.145	3.030	0.710	2.123	3.108	0.704
8	1.662	2.961	0.707	2.163	2.875	0.713	2.400	2.914	0.702
9	2.142	3.145	0.712	1.884	3.061	0.703	2.426	2.967	0.704
10	2.175	3.046	0.707	2.047	3.034	0.706	2.127	2.997	0.705
$\overline{X}$	2.0908	3.0036	0.7069	2.0908	3.0129	0.7064	2.1349	3.0028	0.7039
Target $\overline{\overline{X}}$	2.125	3.000	0.705	2.125	3.000	0.705	2.125	3.000	0.705
$\overline{X}$ − Target $\overline{\overline{X}}$	−0.0342	0.0036	0.0019	−0.0342	0.0129	0.0014	0.0099	0.0028	−0.0011
Target $\overline{s}$	0.234	0.080	0.005	0.234	0.080	0.005	0.234	0.080	0.005

Table 30.7. Cont.

Group number	1			2			3		
Characteristic	A	B	C	A	B	C	A	B	C
$\frac{\bar{X} - \text{Target } \bar{\bar{X}}}{\text{Target } \bar{s}}$	***−0.1462***	0.0450	***0.3800***	***−0.1462***	0.1613	***0.2800***	***0.0423***	0.0350	***−0.2200***
s	0.1756	0.0903	0.0026	0.2240	0.0739	0.0053	0.2537	0.0649	0.0048
s/Target $\bar{s}$	0.7506	***1.1300***	***0.5287***	0.9573	***0.9242***	***1.0549***	***1.0841***	***0.8108***	0.9636
Group number	4			5			6		
Characteristic	A	B	C	A	B	C	A	B	C
1	1.638	3.110	0.710	1.734	2.966	0.712	2.021	2.954	0.707
2	2.442	3.014	0.706	1.485	3.001	0.711	2.171	3.133	0.705
3	2.388	3.034	0.702	2.103	2.989	0.695	2.000	3.046	0.706
4	2.207	3.081	0.705	2.150	3.057	0.712	1.900	2.982	0.706
5	1.946	2.958	0.712	2.217	3.064	0.706	2.398	3.086	0.709
6	2.280	2.971	0.707	2.503	3.125	0.709	2.027	2.954	0.704
7	2.084	2.967	0.703	2.146	3.049	0.701	2.537	2.980	0.704
8	1.777	3.026	0.711	1.919	3.043	0.716	2.160	2.922	0.706
9	2.204	3.018	0.699	2.471	3.011	0.710	1.766	3.040	0.717
10	1.664	3.016	0.695	1.995	3.054	0.695	1.833	2.840	0.715
$\bar{X}$	2.0630	3.0195	0.7050	2.0723	3.0359	0.7067	2.0813	2.9937	0.7079
Target $\bar{\bar{X}}$	2.125	3.000	0.705	2.125	3.000	0.705	2.125	3.000	0.705
$\bar{X}$ − Target $\bar{\bar{X}}$	−0.0620	0.0195	0.000	−0.0527	0.0359	0.0017	−0.0437	−0.0063	0.0029
Target $\bar{s}$	0.234	0.080	0.005	0.234	0.080	0.005	0.234	0.080	0.005
$\frac{\bar{X} - \text{Target } \bar{\bar{X}}}{\text{Target } \bar{s}}$	***−0.2650***	***0.2438***	0.000	***−0.2252***	***0.4488***	0.3400	***−0.1868***	−0.0788	***0.5800***
s	0.2930	0.0485	0.0054	0.3103	0.0455	0.0073	0.2425	0.0850	0.0045
s/Target $\bar{s}$	***1.2522***	***0.6061***	1.0832	1.3261	***0.5690***	***1.4668***	1.0364	***1.0627***	***0.9065***
Group number	7			8			9		
Characteristic	A	B	C	A	B	C	A	B	C
1	2.108	2.880	0.703	2.323	3.074	0.710	2.402	2.994	0.700
2	1.689	2.964	0.709	1.654	2.809	0.704	2.218	3.048	0.698
3	2.309	2.948	0.702	2.289	2.907	0.705	2.069	3.107	0.698
4	2.286	3.031	0.702	2.225	2.978	0.707	2.074	2.976	0.695
5	1.724	2.961	0.703	2.622	3.023	0.710	2.041	2.965	0.704
6	1.856	3.042	0.708	2.130	2.848	0.704	1.943	2.911	0.706
7	2.530	2.986	0.704	1.808	3.050	0.704	2.158	3.037	0.701
8	2.196	3.007	0.706	2.007	3.088	0.699	1.963	3.047	0.713
9	2.654	3.125	0.703	2.137	2.946	0.703	1.920	2.963	0.706
10	2.217	2.925	0.695	2.270	2.891	0.704	1.941	3.004	0.699
$\bar{X}$	2.1569	2.9869	0.7035	2.1465	2.9614	0.7050	2.0729	3.0052	0.7020
Target $\bar{\bar{X}}$	2.125	3.000	0.705	2.125	3.000	0.705	2.125	3.000	0.705
$\bar{X}$ − Target $\bar{\bar{X}}$	0.0319	−0.0131	−0.0015	0.0215	−0.0386	0.000	−0.0521	0.0052	−0.0030
Target $\bar{s}$	0.234	0.080	0.005	0.234	0.080	0.005	0.234	0.080	0.005

Table 30.7. Cont.

Group number	7			8			9		
Characteristic	A	B	C	A	B	C	A	B	C
$\dfrac{\overline{X} - \text{Target } \overline{\overline{X}}}{\text{Target } \overline{s}}$	***0.1363***	−0.1638	***−0.3000***	***0.0919***	***−0.4825***	0.000	−0.2226	***0.0650***	***−0.6000***
s	0.3215	0.0687	0.0039	0.2741	0.0971	0.0033	0.1518	0.0560	0.0053
s/Target $\overline{s}$	***1.3741***	0.8582	***0.7732***	1.1712	***1.2138***	***0.6600***	***0.6487***	0.6997	***1.0583***

Group Short Run $\overline{X}$ *and* s *Chart*

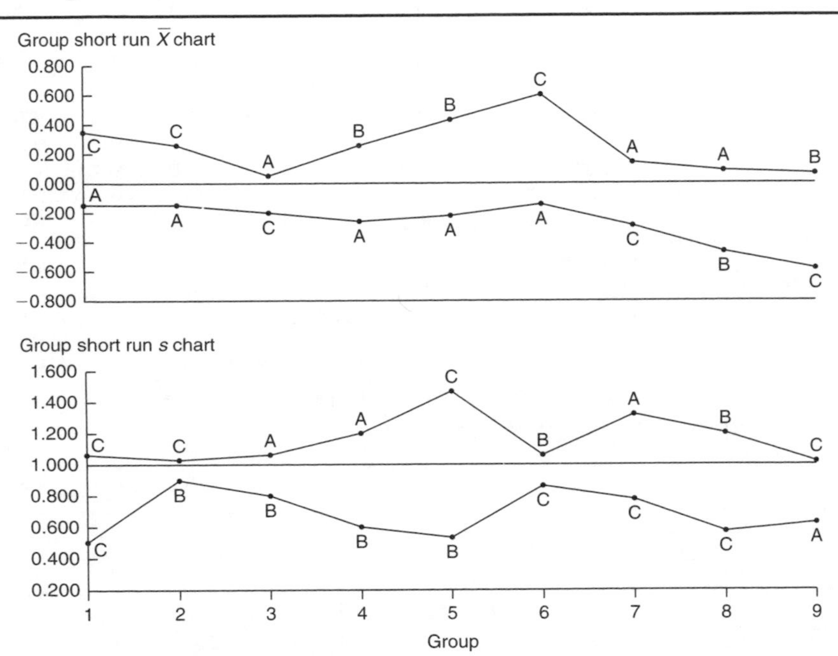

Figure 30.2. Group short run $\overline{X}$ and s chart for the pencil lead example. Three key characteristics are being monitored on the same chart.

Chart Interpretation

Group short run s *chart:* All three characteristics—break force (A), drag (B), and lead diameter (C)—appear to randomly fluctuate in the MAX and MIN positions. This indicates that the initial target $\overline{s}$ values were good estimators for all of the characteristics.

Group short run $\overline{X}$ *chart:* It appears that all three key characteristics are randomly fluctuating in the MAX and MIN positions. This means that the initial target values were good estimators of the actual means for each of the three characteristics.

Recommendations

Group short run s *chart:* Continue using the initial target $\bar{s}$ values for all three characteristics. The charts may look good, but only the capability studies will determine if the characteristics are meeting engineering requirements.

Group short run $\overline{X}$ *chart:* Continue using the initial target $\overline{\overline{X}}$ values. No recalculation is necessary. The process averages appear stable and predictable. Continue to collect data. If the process remains stable, reduce sampling frequency.

Estimating the Process Average

Estimates of the process average should be calculated separately for each characteristic on each part on the group short run charts. The estimate of the process average for break force can be found in Calculation 30.1.

$$\overline{\overline{X}}_A = \frac{\Sigma \overline{X}_A}{k} = \frac{18.9094}{9} = 2.10104$$

Calculation 30.1. Estimate of the process average for characteristic A, break force.

Estimating σ

Estimates of σ are also calculated separately for each characteristic on each part on the group short run charts. Continuing with characteristic A, see Calculations 30.2 and 30.3.

$$\bar{s}_A = \frac{\Sigma s_A}{k} = \frac{2.2465}{9} = 0.24961$$

Calculation 30.2. $\bar{s}$ calculation for characteristic A, break force.

$$\hat{\sigma}_A = \frac{\bar{s}_A}{c_4} = \frac{0.24961}{0.9727} = 0.25662$$

Calculation 30.3. Estimate of the process standard deviation for characteristic A, break force.

Note: To ensure reliable estimates of both the process average and process standard deviation, *k* needs to be at least 20. In this example, *k* is only nine. Therefore, the estimates here and in Table 30.8 are shown only for illustration purposes.

Calculating Process Capability and Performance Ratios

Calculations 30.4, 30.5, and 30.6 show the capability calculations for break force, characteristic A.

$$C_{pA} = \frac{USL_A - LSL_A}{6\hat{\sigma}_A} = \frac{2.80 - 1.20}{6(0.25662)} = \frac{1.6}{1.53972} = 1.04$$

Calculation 30.4. C_p calculation for characteristic A, break force.

$$C_{pk_u A} = \frac{USL_A - \overline{\overline{X}}_A}{3\hat{\sigma}_A} = \frac{2.80 - 2.10104}{3(0.25662)} = \frac{0.69896}{0.76986} = 0.91$$

Calculation 30.5. $C_{pk\ upper}$ for characteristic A, break force.

$$C_{pk_l A} = \frac{\overline{\overline{X}}_A - LSL_A}{3\hat{\sigma}_A} = \frac{2.10104 - 1.20}{3(0.25662)} = \frac{0.90104}{0.76986} = 1.17$$

Calculation 30.6. $C_{pk\ lower}$ calculation for characteristic A, break force.

Group Short Run $\overline{X}$ *and* s *Chart Advantages*

- Graphically illustrates the variation of multiple product or process characteristics relative to each other.
- Characteristics from different parts with different means, different standard deviations, and different units of measure can all be analyzed on the same chart.
- Separates variation due to changes in the average from variation due to changes in the standard deviation.
- Separates variation due to the process from variation that is product specific.

Group Short Run $\overline{X}$ *and* s *Chart Disadvantages*

- No visibility of characteristics that fall between the MAX and MIN plot points
- Cannot detect certain nonrandom conditions because the group charts described here have no control limits
- Lots of calculations

An Additional Comment About the Case

Additional statistics and process capability and performance calculations for key characteristics B and C are shown in Table 30.8.

Table 30.8. Additional statistics and process capability and performance calculations for the drag and diameter key characteristics.

Drag (key B)	Diameter (key C)
$\overline{\overline{X}}_B = 3.00243$	$\overline{\overline{X}}_C = 0.70526$
$\overline{s}_B = 0.07000$	$\overline{s}_C = 0.00471$
$\hat{\sigma}_B = 0.07195$	$\hat{\sigma}_C = 0.00484$
$C_{p\,B} = 2.36$	$C_{p\,C} = 2.06$
$C_{pk_u\,B} = 2.35$	$C_{pk_u\,C} = 1.70$
$C_{pk_l\,B} = 2.37$	$C_{pk_l\,C} = 2.43$

Chapter 31

Conclusion

Unlike the early 1980s, this decade has seen the advent of ISO 9000 and other quality certification systems advocate a resurgence in the use of SPC. This reawakening has led to a renewed emphasis on statistical methods to make processes more productive, capable, and ever-more competitive in a marketplace filled with an overcapacity of suppliers chasing a fixed consumer base. Why the resurgence? Some will point to the ISO 9000 certification requirements and say, "We want SPC because we must become certified." Others will point to their need to bolster organizational competitiveness by improving the consistency and acceptability of their manufactured products. Regardless of the reasons for this SPC renaissance, one thing is certain: it is not easy to use SPC in today's highly complex manufacturing environments where dwindling lot sizes, JIT inventory systems, and an ever-present emphasis on reducing costs pervades managerial thought.

Yet this revived emphasis on statistical methods has brought with it new challenges for SPC practitioners. For example, even if a company is required to use statistical methods to achieve certification,

- How can SPC be used where short production runs and multiple key characteristics are the norm rather than the exception?
- How can SPC be implemented when everyone asks operators to minimize paperwork and maximize their attention to their machines?

- How can a company minimize resources needed to support shop floor SPC while maximizing the usefulness of information that SPC generates?

While we believe that many answers are contained in this text, we also recognize that many new solutions have yet to surface. That is why we welcome suggestions and feedback. Additional ideas, comments, and questions can be addressed to

Innovative Control Charting
P.O. Box 50114
Knoxville, TN 37950-0114
E-mail: innovative@lksys.com

It is our hope that, by using *Innovative Control Charting*, SPC practitioners will be armed with new tools that will allow them to solve most SPC implementation challenges encountered in today's complex manufacturing environment.

Appendix

Factors for Variables Data Control Charts

Table A.1. Factors for variables data control charts. Normal universe factors for computing 3σ control limits.

$\overline{X}$ and range control charts				
n	A_2	D_3	D_4	d_2
1	2.660	0	3.267	1.128
2	1.880	0	3.267	1.128
3	1.023	0	2.574	1.693
4	0.729	0	2.282	2.059
5	0.577	0	2.114	2.326
6	0.483	0	2.004	2.534
7	0.419	0.076	1.924	2.704
8	0.373	0.136	1.864	2.847
9	0.337	0.184	1.816	2.970
10	0.308	0.223	1.777	3.078
11	0.285	0.256	1.744	3.173
12	0.266	0.283	1.717	3.258

Source: American Society for Testing and Materials. Table adapted from ASTM-STP 15D. Copyright ASTM. Reprinted with permission.

Table A.1. Cont.

$\overline{X}$ and s control charts				
n	A_3	B_3	B_4	C_4
6	1.287	0.030	1.970	0.9515
7	1.182	0.118	1.882	0.9594
8	1.099	0.185	1.815	0.9650
9	1.032	0.239	1.761	0.9693
10	0.975	0.284	1.716	0.9727
11	0.927	0.321	1.679	0.9754
12	0.886	0.354	1.646	0.9776
13	0.850	0.382	1.618	0.9794
14	0.817	0.406	1.594	0.9810
15	0.789	0.428	1.572	0.9823
16	0.763	0.448	1.552	0.9835
17	0.739	0.466	1.534	0.9845
18	0.718	0.482	1.518	0.9854
19	0.698	0.497	1.503	0.9862
20	0.680	0.510	1.490	0.9869
21	0.663	0.523	1.477	0.9876
22	0.647	0.534	1.466	0.9882
23	0.633	0.545	1.455	0.9887
24	0.619	0.555	1.445	0.9892
25	0.606	0.565	1.435	0.9896

Given 1:

$$\hat{\sigma} = \frac{\overline{R}}{d_2}$$

Given 2:

$$\hat{\sigma} = \frac{\bar{s}}{c_4}$$

Set the two equations equal to each other:

$$\frac{\overline{R}}{d_2} = \frac{\bar{s}}{c_4}$$

Solve for $\overline{R}$ and replace $\bar{s}$ with s:

$$\text{Target } \overline{R} = \frac{d_2}{c_4} \times s$$

Equation A.1. Derivation of Equation 15.6.

Given 1: When the $C_p = C_{pk} = 1$, then

$$\hat{\sigma} = \frac{1}{6}(\text{USL} - \text{LSL})$$

Given 2:

$$\hat{\sigma} = \frac{\overline{R}}{d_2}$$

Set the two equations equal to each other:

$$\frac{\overline{R}}{d_2} = \frac{1}{6}(\text{USL} - \text{LSL})$$

Solve for $\overline{R}$:

$$\text{Target } \overline{R} = \frac{d_2}{6}(\text{USL} - \text{LSL})$$

Equation A.2. Derivation of Equation 15.7.

Given 1: When $C_{pk} = 1$, then

$$\hat{\sigma} = \frac{1}{3}\left|\text{Spec limit} - \overline{\overline{X}}\right|$$

Given 2:

$$\hat{\sigma} = \frac{\overline{R}}{d_2}$$

Set the two equations equal to each other:

$$\frac{\overline{R}}{d_2} = \frac{1}{3}\left|\text{Spec limit} - \overline{\overline{X}}\right|$$

Solving for $\overline{R}$:

$$\text{Target } \overline{R} = \frac{d_2}{3}\left|\text{Spec limit} - \overline{\overline{X}}\right|$$

Equation A.3. Derivation of Equation 15.8.

Given:

$$\hat{\sigma} = \frac{\bar{s}}{c_4}$$

Substituting Target $\bar{s}$ for $\hat{\sigma}$:

$$\text{Target } \bar{s} = \frac{\bar{s}}{c_4}$$

Substituting s for $\bar{s}$:

$$\text{Target } \bar{s} = \frac{s}{c_4}$$

Equation A.4. Derivation of Equation 15.14.

Given 1: When the $C_p = C_{pk} = 1$, then

$$\hat{\sigma} = \frac{1}{6}(\text{USL} - \text{LSL})$$

Given 2:

$$\hat{\sigma} = \frac{\bar{s}}{c_4}$$

Set the two equations equal to each other:

$$\frac{\bar{s}}{c_4} = \frac{1}{6}(\text{USL} - \text{LSL})$$

Solving for $\bar{s}$:

$$\text{Target } \bar{s} = \frac{c_4}{6}(\text{USL} - \text{LSL})$$

Equation A.5. Derivation of Equation 15.15.

Given 1: When $C_{pk} = 1$, then

$$\hat{\sigma} = \frac{1}{3}\left|\text{Spec limit} - \overline{\overline{X}}\right|$$

Given 2:

$$\hat{\sigma} = \frac{\bar{s}}{c_4}$$

Set the two equations equal to each other:

$$\frac{\bar{s}}{c_4} = \frac{1}{3}\left|\text{Spec limit} - \overline{\overline{X}}\right|$$

Solving for $\bar{s}$:

$$\text{Target } \bar{s} = \frac{c_4}{3}\left|\text{Spec limit} - \overline{\overline{X}}\right|$$

Equation A.6. Derivation of Equation 15.16.

Bibliography

Bothe, Davis R. *SPC for Short Production Runs.* Northville, Mich.: International Quality Institute, Inc., 1988.

Grant, Eugene L. *Statistical Quality Control,* 3d ed. New York: McGraw-Hill, 1964.

Grant, Eugene L., and Richard S. Leavenworth. *Statistical Quality Control,* 7th ed. New York: McGraw-Hill, 1996.

Griffith, Gary K. *Statistical Process Control Methods for Long and Short Runs.* Milwaukee, Wisc.: ASQC Quality Press, 1996.

Western Electric Co. *Statistical Quality Control Handbook.* Indianapolis, Ind.: AT&T, 1956.

Wheeler, Donald J. *Understanding Statistical Process Control,* 2d ed. Knoxville, Tenn.: SPC Press, Inc., 1992.

Index